A GUIDE FOR THE PERPLEXED

ALSO BY E. F. SCHUMACHER

Small Is Beautiful

E. F. SCHUMACHER

A Guide for the Perplexed

HARPER ● PERENNIAL

NEW YORK ● LONDON ● TORONTO ● SYDNEY ● NEW DELHI ● AUCKLAND

HARPER ● PERENNIAL

A hardcover edition of this book was published in 1977 by Harper & Row, Publishers, Inc.

P.S.™ is a trademark of HarperCollins Publishers.

HarperCollins books may be purchased for educational, business, or sales promotional use. For information please e-mail the Special Markets Department at SPsales@harpercollins.com.

FIRST HARPER COLOPHON EDITION PUBLISHED 1978.

REPRINTED IN PERENNIAL 2004 AND REISSUED IN 2015.

ISBN 978-0-06-241481-6 (pbk.)

19   OV/LSC   10 9 8 7 6 5 4

Contents

Nulla est homini causa philoso-
phandi, nisi ut beatus sit.

(Man has no reason to philosophize,
except with a view to happiness.)
                            SAINT AUGUSTINE

1

On Philosophical Maps

On a visit to Leningrad some years ago[1] I consulted a map to find out where I was, but I could not make it out. From where I stood, I could see several enormous churches, yet there was no trace of them on my map. When finally an interpreter came to help me, he said: "We don't show churches on our maps." Contradicting him, I pointed to one that was very clearly marked. "That is a museum," he said, "not what we call a 'living church.' It is only the 'living churches' we don't show."

It then occurred to me that this was not the first time I had been given a map which failed to show many things I could see right in front of my eyes. All through school and university I had been given maps of life and knowledge on which there was hardly a trace of many of the things that I most cared about and that seemed to me to be of the greatest possible importance to the conduct of my life. I remembered that for many years my perplexity had been complete; and no interpreter had come along to help me. It remained complete until I ceased to suspect the sanity of my perceptions and began, instead, to suspect the soundness of the maps.

The maps I was given advised me that virtually all my ancestors, until quite recently, had been rather pathetic illusionists who conducted their lives on the basis of irrational beliefs and absurd superstitions. Even illustrious scientists, like Johannes Kepler or Isaac Newton, apparently spent most of their time and energy on nonsensical studies of nonexisting things. Enormous amounts of hard-earned wealth had been squandered throughout history to the honor and glory of imaginary deities, not only by my European forebears, but by all peoples, in all parts of the world, at all times. Everywhere thousands of seemingly healthy men and women had subjected themselves to utterly meaningless restrictions, like voluntary fasting; tormented themselves by celibacy; wasted their time on pilgrimages, fantastic rituals, reiterated prayers, and so forth; turning their backs on reality—and some do it even in this enlightened age—all for nothing, all out of ignorance and stupidity; none of it to be taken seriously today, except of course as museum pieces. From what a history of error we had emerged! What a history of taking for real what every modern child knew to be totally unreal and imaginary! Our entire past, until quite recently, was today fit only for museums, where people could satisfy their curiosity about the oddity and incompetence of earlier generations. What our ancestors had written, also, was in the main fit only for storage in libraries, where historians and other specialists could study these relics and write books about them, the knowledge of the past being considered interesting and occasionally thrilling but of no particular value for learning to cope with the problems of the present.

All this and many other similar things I was taught at school and university, although not in so many words, not plainly and frankly. It would not do to call a spade a spade. Ancestors had to be treated with respect: they could not help their backwardness; they tried hard and sometimes even got quite near the truth in a haphazard sort of way. Their preoccupation with religion was just one of their many signs of underdevelopment, not surprising in people who had not yet come of age. Even today, of course, there remained some interest in religion, which legitimized that of earlier times. It was still permissible, on suitable occasions, to refer to God the Creator, although every educated person knew that there was not really a God, certainly not one capable of creating anything, and that the things around us had come into existence by a process of mindless evolution, that is, by chance and natural selection. Our ancestors, unfortunately, did not know about evolution, and so they invented all these fanciful myths.

The maps of *real* knowledge, designed for *real* life, showed nothing except things which allegedly could be *proved* to exist. The first principle of the philosophical mapmakers seemed to be "If in doubt, leave it out," or put it into a museum. It occurred to me, however, that the question of *what constitutes proof* was a very subtle and difficult one. Would it not be wiser to turn the principle into its opposite and say: "If in doubt, show it *prominently*"? After all, matters that are beyond doubt are, in a sense, dead; they constitute no challenge to the living.

To accept anything as true means to incur the risk of error. If I limit myself to knowledge that I consider true beyond doubt, I minimize the risk of error, but at the same time I maximize the risk of missing out on what may be the subtlest, most important, and most rewarding things in life. Saint Thomas Aquinas, following Aristotle, taught that "The slenderest knowledge that may be obtained of the highest things is more desirable than the most certain knowledge obtained of lesser things."[2] "Slender" knowledge is here put in opposition to "certain" knowledge, and indicates uncertainty. Maybe it is necessarily so that the *higher* things cannot be known with the same degree of certainty as can the *lesser* things, in which case it would be a very great loss indeed if knowledge were limited to things beyond the possibility of doubt.

The philosophical maps with which I was supplied at school and university did not merely, like the map of Leningrad, fail to show "living churches"; they also failed to show large "unorthodox" sections of both theory and practice in medicine, agriculture, psychology, and the social and political sciences, not to mention art and so-called occult or paranormal phenomena, the mere mention of which was considered to be a sign of mental deficiency. In particular, all the most prominent doctrines shown on the "map" accepted art only as self-expression or as escape from reality. Even in nature there was nothing artistic except by chance, that is to say, even the most beautiful appearances could be fully accounted for—so we were told—by their utility in reproduction, as affecting natural selection. In fact, apart from "museums," the entire map from right to left and from top to bottom was drawn in utilitarian colors: hardly anything was shown as existing unless it could be interpreted as profitable for man's comfort or useful in the universal battle for survival.

Not surprisingly, the more thoroughly acquainted we became with the details of the map, the more we absorbed what it showed and got used to the absence of the things it did not show, the more perplexed, unhappy, and cynical we became. Some of us, however, had experiences similar to that described by Maurice Nicoll:

> Once, in the Greek New Testament class on Sundays, taken by the Head Master, I dared to ask, in spite of my stammering, what some parable meant. The answer was so confused that I actually experienced my first moment of consciousness—that is, I suddenly realised that *no one knew anything* . . . and from that moment I began to think for myself, or rather knew that I could. . . . I remember so clearly this class-room, the high windows constructed so that we could not see out of them, the desks, the platform on which the Head Master sat, his scholarly, thin face, his nervous habits of twitching his mouth and jerking his hands—and suddenly this inner revelation of *knowing that he knew nothing*,—nothing, that is, about anything that really mattered. This was my first inner liberation from the power of external life. From that time, I knew for certain—and that means always by inner individual authentic perception which is the only source of real knowledge—that all my loathing of religion as it was taught me was right.[3]

The maps produced by modern materialistic Scientism leave all the questions that really matter unanswered; more than that, they deny the validity of the questions. The situation was desperate enough in my youth half a century ago; it is even worse now because the ever more rigorous application of the scientific method to all subjects and disciplines has destroyed even the last remnants of ancient wisdom—at least in the Western world. It is being loudly proclaimed *in the name of scientific objectivity* that "values and meanings are nothing but defence mechanisms and reaction formations";[4] that man is "nothing but a complex biochemical mechanism powered by a combustion system which energises computers with prodigious storage facilities for retaining encoded information."[5] Sigmund Freud even assured us that "this alone I know with certainty, namely that men's value judgments are guided absolutely by their desire for happiness, and are therefore merely an attempt to bolster up their illusions by arguments."[6]

How is anyone to resist the pressure of such statements, made in the name of objective science, unless, like Maurice Nicoll, he suddenly receives "this inner revelation of knowing" that men who say such things, however learned they may be, *know nothing about anything that really matters?* People are asking for bread and they are being given stones. They beg for advice about what they should do "to be saved," and they are told that the idea of salvation has no intelligible content and is nothing but an infantile neurosis. They long for guidance about how to live as responsible human beings, and they are told that they are machines, like computers, without free will and therefore without responsibility.

"The present danger," says Viktor E. Frankl, a psychiatrist of unshakable sanity, "does not really lie in the loss of universality on the part of the scientist, but rather in his pretence and claim of totality. . . . What we have to deplore therefore is not so much the fact that *scientists are specialising,* but rather the fact that *specialists are generalising.*" After many centuries of theological imperialism, we have now had three centuries of an ever more aggressive "scientific imperialism," and the result is a degree of bewilderment and disorientation, particularly among the young, which can at any moment lead to the collapse of our civilization. "The true nihilism of today," says Dr. Frankl, "is reductionism. . . . Contemporary nihilism no longer brandishes the word nothingness; today nihilism is camouflaged as *noth-*

ing-but-ness. Human phenomena are thus turned into mere epiphenomena."[7]

Yet they remain *our reality,* everything we are and everything we become. In this life we find ourselves as in a strange country. Ortega y Gasset once remarked that "Life is fired at us point-blank." We cannot say: "Hold it! I am not quite ready. Wait until I have sorted things out." Decisions have to be taken that we are not ready for; aims have to be chosen that we cannot see clearly. This is very strange and, on the face of it, quite irrational. Human beings, it seems, are insufficiently "programmed." Not only are they utterly helpless when they are born and remain so for a long time; even when fully grown, they do not move and act with the sure-footedness of animals. They hesitate, doubt, change their minds, run hither and thither, uncertain not simply of how to get what they want but above all of *what* they want.

Questions like "What should I do?" or "What must I do to be saved?" are strange questions because they relate to *ends,* not simply to means. No technical answer will do, such as "Tell me precisely what you want and I shall tell you how to get it." The whole point is that I do not know what I want. Maybe all I want is to be happy. But the answer "Tell me what you need for happiness, and I shall then be able to advise you what to do"— this answer, again, will not do, because I do not know what I need for happiness. Perhaps someone says: "For happiness you need wisdom"—but what is wisdom? Or: "For happiness you need the truth that makes you free"—but what is the truth that makes us free? Who will tell me where I can find it? Who can guide me to it or at least point out the direction in which I have to proceed?

In this book, we shall look at the world and try and see it whole. To do this is sometimes called to philosophize, and philosophy has been defined as the love of, and seeking after, wisdom. Socrates said: "Wonder is the feeling of a philosopher, and philosophy begins with wonder." He also said: "No god is a philosopher or seeker after wisdom for he is wise already. Neither do the ignorant seek after wisdom; for herein is the evil of ignorance, that he who is neither good nor wise is nevertheless satisfied with himself."[8]

One way of looking at the world as a whole is by means of a map, that is to say, some sort of a plan or outline that shows where various things are to be found—not all things, of course, for that would make the map as big as the world, but the things that are most prominent, most important for orientation—outstanding landmarks, as it were, which you cannot miss, or if you do miss them, you will be left in total perplexity.

The most important part of any inquiry or exploration is its beginning. As has often been pointed out, if one makes a false or superficial beginning, no matter how rigorous the methods followed during the succeeding investigation, they will never remedy the initial error.[9]

Mapmaking is an empirical art that employs a high degree of abstraction but nonetheless clings to reality with something akin to self-abandonment. Its motto, in a sense, is "Accept everything; reject nothing." If something is *there*, if it has any kind of existence, if people notice it and are interested in it, it must be indicated on the map, in its proper place. Mapmaking is not the whole of philosophy, just as a map or guidebook is not the whole of geography. It is simply a beginning—the very beginning which is at present lacking, when people ask: "What does it all mean?" or "What am I supposed to do with my life?"

My map or guidebook is constructed on the recognition of four Great Truths—or landmarks—which are so prominent, so all-pervading, that you can see them wherever you happen to be. If you know them well, you can always find your location by them, and if you cannot recognize them, you are lost.

The guidebook, it might be said, is about how "Man lives in the world." This simple statement indicates that we shall need to study

1. "The world";
2. "Man"—his equipment to meet the world;
3. His way of learning about the world; and
4. What it means to "live" in this world.

The Great Truth about the world is that it is a hierarchic structure of at least four great "Levels of Being."

The Great Truth about man's equipment to meet the world is the principle of "adequateness" *(adaequatio)*.

The Great Truth about man's learning concerns the "Four Fields of Knowledge."

The Great Truth about living in this life, living in this world, relates to the distinction between two types of problem, "convergent" and "divergent."

A map or guidebook—let this be understood as clearly as possible—does not "solve" problems and does not "explain" mysteries; it merely helps to identify them. Thereafter, everybody's task is as defined by the last words spoken by the Buddha:

Work out your salvation with diligence.

For this purpose, according to the precepts of the Tibetan teachers,

> A philosophy comprehensive enough to embrace the whole of knowledge is indispensable.
>
> A system of meditation which will produce the power of concentrating the mind on anything whatsoever is indispensable.
>
> An art of living which will enable one to utilise each activity (of body, speech and mind) as an aid on the Path is indispensable.[10]

II

The more recent philosophers of Europe have seldom been faithful mapmakers. Descartes (1596–1650), for instance, to whom modern philosophy owes so much, approached his self-set task in quite a different way. "Those who seek the direct road to truth," he said, "should not bother with any object of which they cannot have a certainty equal to the demonstrations of arithmetic and geometry." Only such objects should engage our attention "to the sure and indubitable knowledge of which our mental powers seem to be adequate."[11]

Descartes, the father of modern rationalism, insisted that "We should never allow ourselves to be persuaded excepting by the evidence of our Reason," and he stressed particularly that he spoke "of our Reason and not of our imagination nor of our senses."[12] The method of reason is to *"reduce involved and obscure propositions step by step to those that are simpler, and then, starting with the intuitive apprehension of all those that are absolutely simple, attempt to ascend to the knowledge of all others by precisely similar steps."*[13] This is a program conceived by a mind both powerful and frighteningly narrow, whose narrowness is further demonstrated by the rule:

> If in the matters to be examined we come to a step in the series of which our understanding is not sufficiently well able to have an intuitive cognition, we must stop short there. We must make no attempt to examine what follows; thus we shall spare ourselves superfluous labour.[14]

Descartes limits his interest to knowledge and ideas that are precise and certain beyond any possibility of doubt, because his primary interest is that we should become *"masters and possessors of nature."* Nothing can be precise unless it can be quantified in one way or another. As Jacques Maritain comments:

> The mathematical knowledge of nature, for Descartes, is not what it is in reality, a certain interpretation of phenomena . . . which does not answer questions bearing upon the first principles of things. This knowledge is, for him, the revelation of the very essence of things. These are analysed exhaustively by geometric extension and local movement. The whole of physics, that is, the whole of the philosophy of nature, is nothing but geometry.
>
> Thus Cartesian evidence goes straight to mechanism. It mechanises nature; it does violence to it; it annihilates everything which causes things to symbolise with the spirit, to partake of the genius of the Creator, to speak to us. The universe becomes dumb.[15]

There is no guarantee that the world is made in such a way that indubitable truth is the whole truth. *Whose* truth, *whose* understanding, would it be? That of man. Of any man? Are all men "adequate" to grasp all truth? As Descartes has demonstrated, the mind of man can doubt everything it cannot grasp with ease, and some men are more prone to doubt than others.

Descartes broke with tradition, made a clean sweep, and undertook to start afresh, to find out everything for himself. This kind of arrogance became the "style" of European philosophy. "Every modern philosopher," as Maritain remarks, "is a Cartesian in the sense that he looks upon himself as starting off in the absolute, and as having the mission of bringing men a new conception of the world."[16]

The alleged fact that philosophy "had been cultivated for many centuries by the best minds that have ever lived and that nevertheless no single thing is to be found in it which is not a subject of dispute and in consequence is not dubious"[17] led Descartes to what amounted to a "withdrawal from wisdom" and exclusive concentration on knowledge as firm and indubitable as mathematics and geometry. Francis Bacon (1561–1626) had already pleaded in a similar vein. Skepticism, a form of defeatism in philosophy, became the main current of European philosophy, which insisted, not without plausibility, that the reach of the human mind was strictly limited and that there was no point in taking any interest in matters beyond its capacity. While traditional wisdom had considered the human mind as weak but *open-ended*—that is, capable of reaching beyond itself toward higher and higher levels—the new thinking took it as axiomatic that the mind's reach had fixed and narrow limits, which could be clearly determined, while within these limits it possessed virtually unlimited powers.

From the point of view of philosophical mapmaking, this meant a very great impoverishment: entire regions of human interest, which had engaged the most intense efforts of earlier generations, simply ceased to appear on the maps. But there was an even more significant withdrawal and impoverishment: While traditional wisdom had always presented the world as a three-dimensional structure (as symbolized by the cross), where it was not only meaningful but essential to distinguish always and everywhere between "higher" and "lower" things and Levels of Being, the new thinking strove with determination, not to say fanaticism, to get rid of the *vertical dimension*. How could one obtain clear and precise ideas about such qualitative notions as "higher" or "lower"? Was it not reason's most urgent task to replace them with quantitative measurements?

But perhaps the "mathematicism" of Descartes had gone too far; so Immanuel Kant (1724–1804) set out to make a new start. Yet as Etienne Gilson, the incomparable master of the history of philosophy, remarks:

> Kant was not shifting from mathematics to philosophy, but from mathematics to physics. As Kant himself immediately concluded: "The true method of metaphysics is fundamentally the same as that which Newton has introduced into natural science, and which has there yielded such fruitful results.". . . *The Critique of Pure Reason* is a masterly description of what the structure of the human mind should be, in order to account for the existence of a Newtonian conception of nature, and assuming that conception to be true to reality. Nothing can show more clearly the essential weakness of physicism as a philosophical method.[18]

Neither mathematics nor physics can entertain the qualitative notion of "higher" or "lower." So the *vertical dimension* disappeared from the philosophical maps, which henceforth concentrated on somewhat farfetched problems, such as "Do other people exist?" or "How can I know anything at all?" or "Do other people have experiences analogous to mine?" Thus the maps ceased to be of any help to people in the awesome task of picking their way through life.

The proper task of philosophy was formulated by Etienne Gilson as follows:

> It is its permanent duty to order and to regulate an ever wider area of scientific knowledge, and to judge ever more complex problems of human conduct; it is its never-ended task to keep the old sciences in their natural limits, to assign their places, and their limits, to new sciences; last, not least, to keep all human activities, however changing their circumstances, under the sway of the same reason by which alone man remains the judge of his own works and, after God, the master of his own destiny.[19]

III

The loss of the *vertical dimension* meant that it was no longer possible to give an answer, other than a utilitarian one, to the question "What am I to do with my life?" The answer could be more individualistic-selfish or more social-unselfish, but it could not help being utilitarian: either "Make yourself as comfortable as you can" or "Work for the greatest happiness of the greatest number." Nor was it possible to define the nature of man other than as that of an animal. A "higher" animal? Yes, perhaps; but only in some respects. In certain respects other animals could be described as "higher" than man, and so it would be best to avoid nebulous terms like "higher" or "lower," unless one spoke in strictly *evolutionary* terms. In the context of evolution, "higher" could generally be associated with "later," and since man was undoubtedly a latecomer, he could be thought of as standing at the top of the evolutionary ladder.

None of this leads to a helpful answer to the question "What am I to do with my life?" Pascal (1623–1662) had said: "Man wishes to be happy and exists only to be happy and cannot wish not to be happy,"[20] but the new thinking of the philosophers insisted, with Kant, that man "never can say definitely and consistently what it is that he really wishes," nor can he "determine with certainty what would make him truly happy; because to do so he would need to be omniscient."[21] Traditional wisdom had a reassuringly plain answer: Man's happiness is to move *higher*, to develop his *highest* facilities, to gain knowledge of the *highest* things and, if possible, to "see God." If he moves *lower*, develops only his *lower* faculties, which he shares with the animals, then he makes himself deeply unhappy, even to the point of despair.

With imperturbable certainty Saint Thomas Aquinas (1225–1274) argued:

No man tends to do a thing by his desire and endeavour unless it be previously known to him. Wherefore since man is directed by divine providence *to a higher good than human frailty can attain* in the present life ... it was necessary for his mind to be bidden to *something higher* than those things to which our reason can reach in the present life, *so that he might learn to aspire*, and by his endeavours to tend to *something surpassing the whole state of the present life.* . . . It was with this motive that the philosophers, in order to wean men from sensible pleasures to virtue, took care to show that there are other goods of greater account than those which appeal to the senses, the taste of which things affords much greater delight to those who devote themselves to active or contemplative virtues.[22]

These teachings, which are the traditional wisdom of all peoples in all parts of the world, have become virtually incomprehensible to modern man, although he, too, desires nothing more than somehow to be able to rise above "the whole state of the present life." He hopes to do so by growing rich, by moving around at ever-increasing speed, by traveling to the moon and into space. It is worth listening again to Saint Thomas:

There is a desire in man, common to him and other animals, namely the desire for *the enjoyment of pleasure:* and this men pursue especially by leading a voluptuous life, and through lack of moderation become intemperate and incontinent. Now in that vision [the divine vision] there is the most perfect pleasure, all the more perfect than sensuous pleasure as the intellect is above the senses; as the good in which we shall delight surpasses all sensible good, is more penetrating, and more continuously delightful; and as that pleasure is freer from all alloy of sorrow or trouble of anxiety. . . .

In this life there is nothing so like this ultimate and perfect happiness as the life of those who contemplate the truth, as far as possible here below. Hence the philosophers who were unable to obtain full knowledge of that final beatitude, placed man's ultimate happiness in that contemplation which is possible during this life. For this reason too, Holy Writ commends the contemplative rather than other forms of life, when our Lord said (Luke X. 42): *Mary hath chosen the better part*, namely the contemplation of truth, *which shall not be taken from her.* For contemplation of truth begins in this life, but will be consummated in the life to come: while the active and civic life does not transcend the limits of this life.[23]

Most modern readers will be reluctant to believe that perfect happiness is attainable by methods of which their modern world knows nothing. However, belief or disbelief is not the matter at issue here. The point is that without the qualitative concepts of "higher" and "lower" it is impossible even to think of guidelines for living which lead beyond individual or collective utilitarianism and selfishness.

The ability to see the Great Truth of the hierarchic structure of the world, which makes it possible to distinguish between *higher and lower Levels of Being,* is one of the indispensable conditions of understanding. Without it, it is not possible to find out every thing's proper and legitimate place. Everything, everywhere, can be understood only when its *Level of Being* is fully taken into account. Many things which are true at a low Level of Being become absurd at a higher level, and of course vice versa.

We therefore now turn to a study of the hierarchic structure of the world.

2

Levels of Being

Our task is to look at the world and see it whole.

We see what our ancestors have always seen: a great Chain of Being which seems to divide naturally into four sections— four "kingdoms," as they used to be called: mineral, plant, animal, and human. This "was, in fact, until not much more than a century ago, probably the most widely familiar conception of the general *scheme* of things, of the constitutive pattern of the universe."[1] The Chain of Being can be seen as extending downward from the Highest to the lowest, or it can be seen as extending upward from the lowest to the Highest. The ancient view begins with the Divine and sees the downward Chain of Being as moving an ever-increasing distance from the Center, with a progressive loss of qualities. The modern view, largely influenced by the doctrine of evolution, tends to start with inanimate matter and to consider man the last link of the chain, as having evolved the widest range of useful qualities. For our purposes here, the direction of looking—upward or downward—is unimportant, and, in line with modern habits of thought, we shall start at the lowest level, the mineral kingdom, and consider the successive gain of qualities or *powers* as we move to the higher levels.

No one has any difficulty recognizing the astonishing and mysterious difference between a living plant and one that has died and has thus fallen to the lowest Level of Being, inanimate matter. What is this *power* that has been lost? We call it "life." Scientists tell us that we must not talk of a "life force" because no such force has ever been found to exist. Yet the *difference* between alive and dead exists. We could call it "*x*," to indicate something that is there to be noticed and studied but that cannot be explained. If we call the mineral level "*m*," we can call the plant level $m + x$. This factor x is obviously worthy of our closest attention, particularly since we are able to destroy it, although it is completely outside our ability to create it. Even if somebody could provide us with a recipe, a set of instructions, for creating life out of lifeless matter, the mysterious character of x would remain, and we would never cease to marvel that something that could do nothing is now able to extract nourishment from its environment, grow, and reproduce itself, "true to form," as it were. There is nothing in the laws, concepts, and formulae of physics and chemistry to explain or even to describe such powers. X is something quite new and additional, and the more deeply we contemplate it, the clearer it becomes that we are faced here with what might be called an *ontological discontinuity* or, more simply, a jump in the Level of Being.

From plant to animal, there is a similar jump, a similar addition of powers, which enable the typical, fully developed animal to do things that are totally outside the range of possibilities of the typical, fully developed plant. These powers, again, are mysterious and, strictly speaking, nameless. We can refer to them by the letter "*y*," which will be the safest course, because any word label we might attach to them could lead people to think that such a designation was not merely a hint as to their nature but an adequate description. However, since we cannot talk without words, I shall attach to these mysterious powers the label *consciousness*. It is easy to recognize consciousness in a dog, a cat, or a horse, if only because they can be knocked unconscious: the processes of life continue as in a plant, although the animal has lost its peculiar powers.

If the plant, in our terminology, can be called $m + x$, the

animal has to be described as $m + x + y$. Again, the new factor
"y" is worthy of our closest attention; we are able to destroy but
not to create it. Anything that we can destroy but are unable
to make is, in a sense, sacred, and all our "explanations" of it do
not really explain anything. Again we can say that y is something quite new and additional when compared with the level
"plant"—another *ontological discontinuity*, another jump in
the Level of Being.

Moving from the animal to the human level, who would
seriously deny the addition, again, of new powers? What precisely they are has become a matter of controversy in modern
times, but the fact that man is able to do—and is doing—innumerable things which lie totally outside the range of possibilities of even the most highly developed animals cannot be disputed and has never been denied. Man has powers of life like
the plant, powers of consciousness like the animal, and evidently something more: the mysterious power "z". What is it?
How can it be defined? What can it be called? This power z
has undoubtedly a great deal to do with the fact that man is
not only able to think but is also *able to be aware of his thinking*. Consciousness and intelligence, as it were, recoil upon
themselves. There is not merely a conscious being, but a being
capable of being conscious of its consciousness; not merely a
thinker, but a thinker capable of watching and studying his
own thinking. There is something able to say "I" and *to direct
consciousness* in accordance with its own purposes, a master
or controller, a power at a higher level than consciousness itself. This power z, consciousness recoiling upon itself, opens
up unlimited possibilities of purposeful learning, investigating,
exploring, and of formulating and accumulating knowledge.
What shall we call it? As it is necessary to have word labels, I
shall call it *self-awareness*. We must, however, take great care
always to remember that such a word label is merely (to use a
Buddhist phrase) "a finger pointing to the moon." The
"moon" itself remains highly mysterious and needs to be studied with the greatest patience and perseverance if we want to
understand anything about man's position in the Universe.

Our initial review of the four great Levels of Being can be summed up as follows:

Man can be written $m + x + y + z$
Animal can be written $m + x + y$
Plant can be written $m + x$
Mineral can be written m

Only m is visible; x, y, and z are invisible, and they are extremely difficult to grasp, although their effects are matters of everyday experience.

If, instead of taking "minerals" as our base line and reaching the higher Levels of Being by the addition of powers, we start with the highest level directly known to us—man—we can reach the lower Levels of Being by the progressive subtraction of powers. We can then say:

Man can be written M
Animal can be written $M - z$
Plant can be written $M - z - y$
Mineral can be written $M - z - y - x$

Such a downward scheme is easier for us to understand than the upward one, simply because it is closer to our practical experience. We know that all three factors—x, y, and z—can weaken and die away; we can in fact deliberately destroy them. Self-awareness can disappear while consciousness continues; consciousness can disappear while life continues; and life can disappear leaving an inanimate body behind. We can observe, and in a sense *feel*, the process of diminution to the point of the apparently total disappearance of self-awareness, consciousness, and life. But it is outside our power to give life to inanimate matter, to give consciousness to living matter, and finally to add the power of self-awareness to conscious beings.

What we can do ourselves, we can, in a sense, understand; what we cannot do at all, we cannot understand—not even "in a sense." Evolution as a process of the spontaneous, accidental emergence of the powers of life, consciousness, and self-awareness, out of inanimate matter, is totally incomprehensible.

For our purposes, however, there is no need to enter into such speculations at this stage. We hold fast to what we can see and experience: the Universe is as a great hierarchic structure of four markedly different Levels of Being. Each level is obviously a broad band, allowing for higher and lower beings within each band, and the precise determination of where a lower band ends and a higher band begins may sometimes be a matter of difficulty and dispute. The existence of the four kingdoms, however, is not put into question by the fact that some of the frontiers are occasionally disputed.

Physics and chemistry deal with the lowest level, "minerals." At this level, x, y, and z—life, consciousness, and self-awareness —do not exist (or, in any case, are totally inoperative and therefore cannot be noticed). Physics and chemistry can tell us nothing, *absolutely nothing*, about them. These sciences possess no concepts relating to such powers and are incapable of describing their effects. Where there is life, there is form, *Gestalt*, which reproduces itself over and over again from seed or similar beginnings which do not possess this *Gestalt* but develop it in the process of growth. Nothing comparable is to be found in physics or chemistry.

To say that life is nothing but a property of certain peculiar combinations of atoms is like saying that Shakespeare's *Hamlet* is nothing but a property of a peculiar combination of letters. The truth is that the peculiar combination of letters is nothing but a property of Shakespeare's *Hamlet*. The French or German versions of the play "own" different combinations of letters.

The extraordinary thing about the modern "life sciences" is that they hardly ever deal with *life as such*, the factor x, but devote infinite attention to the study and analysis of the physicochemical body that is life's carrier. It may well be that modern science has no method for coming to grips with *life as such*. If this is so, let it be frankly admitted; there is no excuse for the pretense that life is nothing but physics and chemistry.

Nor is there any excuse for the pretense that consciousness is nothing but a property of life. To describe an animal as a physicochemical system of extreme complexity is no doubt perfectly correct, except that it misses out on the "animalness" of the animal. Some zoologists, at least, have advanced beyond this level of erudite absurdity and have developed an ability to see in animals more than complex machines. Their influence, however, is as yet deplorably small, and with the increasing "rationalization" of the modern life-style, more and more animals are being treated as if they really were nothing but "animal machines." (This is a very telling example of how philosophical theories, no matter how absurd and offensive to common sense, tend to become, after a while, "normal practice" in everyday life.)

All the "humanities," as distinct from the natural sciences, deal in one way or another with factor y—consciousness. But a distinction between consciousness (y) and self-awareness (z) is seldom drawn. As a result, modern thinking has become increasingly uncertain whether or not there is any "real" difference between animal and man. A great deal of study of the behavior of animals is being undertaken for the purpose of understanding the nature of man. This is analogous to studying physics with the hope of learning something about life (x). Naturally, since man, as it were, *contains* the three lower Levels of Being, certain things about him can be elucidated by studying minerals, plants, and animals—in fact, everything can be learned about him *except that which makes him human*. All the four constituent elements of the human person—m, x, y, and z—deserve study, but there can be little doubt about their relative importance in terms of *knowledge for the conduct of our lives*.

This importance increases in the order given above, and so do the difficulty and uncertainty experienced by modern humanity. Is there really anything beyond the world of matter, of molecules and atoms and electrons and innumerable other small particles, the ever more complex combinations of which allegedly account for simply everything, from the crudest to the most sublime? Why talk about fundamental differences, "jumps" in the Chain of Being, or "ontological discontinuities" when all we can be really sure of are *differences in degree?* It is not necessary for us to battle over the question whether the palpable and overwhelmingly obvious differences between the four great Levels of Being are better seen as differences in kind or differences in degree. What has to be fully understood is that there are differences in kind, and not simply in degree, between the *powers* of life, consciousness, and self-awareness. Traces of these powers may already exist at the lower levels, although not noticeable (or not yet noticed) by man. Or maybe they are infused, so to speak, on appropriate occasions from "another world." It is not essential for us to have theories about their origin, provided we recognize their quality and, in so doing, never fail to remember that they are beyond anything our own intelligence enables us to create.

It is not unduly difficult to appreciate the difference between what is alive and what is lifeless; it is more difficult to distinguish consciousness from life; and to realize, experience, and appreciate the difference between self-awareness and consciousness (that is, between z and y) is hard indeed. The reason for the difficulty is not far to seek: While the higher comprises and therefore in a sense understands the lower, no being can understand anything higher than itself. A human being can indeed strain and stretch toward the higher and induce a process of growth through adoration, awe, wonder, admiration, and imitation, and by attaining a higher level expand its understanding (and this is a subject that will occupy us extensively later on). But people within whom the power of self-awareness (z) is poorly developed cannot grasp it as a separate power and tend to take it as *nothing but* a slight extension of consciousness (y). Hence we are given a large number of definitions of man which make him out to be *nothing but* an exceptionally intelligent animal with a measurably larger brain, or a tool-making animal, or a political animal, or an unfinished animal, or simply a naked ape.

No doubt, people who use these terms cheerfully include themselves in their definitions—and may have some reason for doing so. For others, they sound merely inane, like defining a dog as a barking plant or a running cabbage. Nothing is more conducive to the brutalization of the modern world than the launching, in the name of science, of wrongful and degrading definitions of man, such as "the naked ape." What could one expect of such a creature, of other "naked apes," or, indeed, of oneself? When people speak of animals as "animal machines," they soon start treating them accordingly, and when they think of people as naked apes, all doors are opened to the free entry of bestiality.

"What a piece of work is a man! how noble in reason! how infinite in faculty!" Because of the power of self-awareness (z), his faculties are indeed infinite; they are not narrowly determined, confined, or "programmed" as one says today. Werner Jaeger expressed a profound truth in the statement that once a human potentiality is realized, it exists. It is the greatest human achievements that define man, not any average behavior or performance, and certainly not anything that can be derived from the observation of animals. "All men cannot be outstanding," says Catherine Roberts. "Yet all men, through knowledge of superior humanness, could know what it means to be a human being and that, as such, they too have a contribution to make. It is magnificent to become as human as one is able. And it requires no help from science. In addition, the very act of realising one's potentialities might constitute an advance over what has gone before."[2]

This "open-endedness" is the wonderful result of the specifically human powers of self-awareness (z), which, as distinct from the powers of life and consciousness, have nothing automatic or mechanical about them. The powers of self-awareness are *essentially* a limitless potentiality rather than an actuality. They have to be developed and "realized" by each human individual if he is to become truly human, that is to say, a *person*.

I said earlier on that man can be written

$$m + x + y + z.$$

These four elements form a sequence of increasing rarity and vulnerability. Matter (m) cannot be destroyed; to kill a body means to deprive it of x, y, and z, and the inanimate matter remains; it "returns" to the earth. Compared with inanimate matter, life is rare and precarious; in turn, compared with the ubiquitousness and tenacity of life, consciousness is even rarer and more vulnerable. Self-awareness is the rarest power of all, precious and vulnerable to the highest degree, the supreme and generally fleeting achievement of a person, present one moment and all too easily gone the next. The study of this factor z has in all ages—except the present—been the primary concern of mankind. How is it possible to study something so vulnerable and fleeting? How is it possible to study that which does the studying? How, indeed, can I study the "I" that employs the very consciousness needed for the study? These questions will occupy us in a later part of this book. Before we can turn to them directly, we shall do well to take a closer look at the four great Levels of Being: how the intervention of additional powers introduces *essential* changes, even though similarities and "correspondences" remain.

Matter (m), life (x), consciousness (y), self-awareness (z)—these four elements are ontologically—that is, in their fundamental nature—different, incomparable, incommensurable, and discontinuous. Only one of them is directly accessible to objective, scientific observation by means of our five senses. The other three are none the less known to us because we ourselves, every one of us, can verify their existence from our own inner experience.

We never find life except as living matter; we never find consciousness except as conscious living matter; and we never find self-awareness except as self-aware, conscious, living matter. The ontological differences between these four elements are analogous to the discontinuity of dimensions. A line is one-dimensional, and no elaboration of a line, no subtlety in its construction, and no complexity can ever turn it into a surface. Equally, no elaboration of a two-dimensional surface, no increase in complexity, subtlety, or size, can ever turn it into a solid. Existence in the physical world we know is attained only by three-dimensional beings. One- or two-dimensional things exist only in our minds. Analogically speaking, it might be said that only man has "real" existence in this world insofar as he alone possesses the "three dimensions" of life, consciousness, and self-awareness. In this sense, animals, with only two dimensions—life and consciousness—have but a shadowy existence, and plants, lacking the dimensions of self-awareness and consciousness, relate to a human being as a line relates to a solid. In terms of this analogy, matter, lacking the three "invisible dimensions," has no more reality than a geometrical point.

This analogy, which may seem farfetched from a logical point of view, points to an inescapable *existential* truth: The most "real" world we live in is that of our fellow human beings. Without them we should experience a sense of enormous emptiness; we could hardly be human ourselves, for we are made or marred by our relations with other people. The company of animals could console us only because, and to the extent to which, they were reminders, even caricatures, of human beings. A world without fellow human beings would be an eerie and unreal place of banishment; with neither fellow humans nor animals the world would be a dreadful wasteland, no matter how luscious its vegetation. To call it one-dimensional would not seem to be an exaggeration. Human existence in a totally inanimate environment, if it were possible, would be total emptiness, total despair. It may seem absurd to pursue such a line of thought, but it is surely not so absurd as a view which counts as "real" only inanimate matter and treats as "unreal," "subjective," and therefore scientifically nonexistent the invisible dimensions of life, consciousness, and self-awareness.

A simple inspection of the four great Levels of Being has led us to the recognition of their four "elements"—matter, life, consciousness, and self-awareness. It is this recognition that matters, not the precise association of the four elements with the four Levels of Being. If the natural scientists should come and tell us that there are some beings they call animals in whom no trace of consciousness can be detected, it would not be for us to argue with them. Recognition is one thing; identification quite another. For us, only recognition is important, and we are entitled to choose for our purposes typical and fully developed specimens from each Level of Being. If they manifest and demonstrate most clearly the "invisible dimensions" of life, consciousness, and self-awareness, this demonstration is not nullified or invalidated by any difficulty of classification in other cases.

Once we have recognized the ontological gaps and discontinuities that separate the four "elements"—m, x, y, z—from one another, we know also that there can exist no "links" or "transitional forms": Life is either present or absent; there cannot be a half-presence; and the same goes for consciousness and self-awareness. Difficulties of identification are often increased by the fact that the lower level appears to present a kind of mimicry or counterfeit of the higher, just as an animated puppet can at times be mistaken for a living person, or a two-dimensional picture can look like three-dimensional reality. But neither difficulties of identification and demarcation nor possibilities of deception and error can be used as arguments against the existence of the four great Levels of Being, exhibiting the four "elements" we have called Matter, Life, Consciousness, and Self-awareness. These four "elements" are four irreducible mysteries, which need to be most carefully observed and studied, but which cannot be explained, let alone "explained away."

In a hierarchic structure, the higher does not merely possess powers that are additional to and exceed those possessed by the lower; it also has power *over* the lower: it has the power to organize the lower and use it for its own purposes. Living beings can organize and utilize inanimate matter, conscious beings can utilize life, and self-aware beings can utilize consciousness. Are there powers that are higher than self-awareness? Are there Levels of Being above the human? At this stage in our investigation we need do no more than register the fact that the great majority of mankind throughout its known history, until very recently, has been unshakenly convinced that the Chain of Being extends upward beyond man. This universal conviction of mankind is impressive for both its duration and its intensity. Those individuals of the past whom we still consider the wisest and greatest not only shared this belief but considered it of all truths the most important and the most profound.

3

Progressions

The four great Levels of Being exhibit certain characteristics in a manner which I shall call *progressions*. Perhaps the most striking progression is the movement from Passivity to Activity. At the lowest level, that of "minerals" or inanimate matter, there is pure passivity. A stone is wholly passive, a pure object, totally dependent on circumstances and "contingent." It can do nothing, organize nothing, utilize nothing. Even radioactive material is passive.

A plant is mainly, but not totally, passive; it is not a pure object; it has a certain, limited ability of adaptation to changing circumstances: it grows toward the light and extends its roots toward moisture and nutrients in the soil. A plant is to a small extent a *subject*, with its own power of doing, organizing, and utilizing. It can even be said that there is an intimation of active intelligence in plants—not, of course, as active as that of animals.

At the level of "animal," through the appearance of consciousness, there is a striking shift from passivity to activity. The processes of life are speeded up; activity becomes more autonomous, as evidenced by free and often purposeful movement—not merely a gradual turning toward light but swift action to obtain food or escape danger. The power of doing, organizing, and utilizing is immeasurably extended; there is evidence of an "inner life," of happiness and unhappiness, confidence, fear, expectation, disappointment, and so forth. Any being with an inner life cannot be a mere object: it is a subject itself, capable even of treating other beings as mere objects, as the cat treats the mouse.

At the human level, there is a subject that says "I"—a person: another marked change from passivity to activity, from object to subject. To treat a person as if he or she were a mere object is a perversity, not to say a crime. No matter how weighed down and enslaved by circumstances a person may be, there always exists the possibility of self-assertion and rising above circumstances. Man can achieve a measure of control over his environment and thereby his life, utilizing things around him for his own purposes. There is no definable limit to his possibilities, even though he everywhere encounters practical limitations which he has to recognize and respect.

This progressive movement from passivity to activity, which we observe in the four Levels of Being, is indeed striking, but it is not complete. A large *weight* of passivity remains even in the most sovereign and autonomous human person; while he is undoubtedly a *subject*, he remains in many respects an *object* —dependent, contingent, pushed around by circumstances. Aware of this, mankind has always used its imagination, or its intuitive powers, to complete the process, to extrapolate (as we might say today) the observed curve to its completion. Thus was conceived a Being, wholly active, wholly sovereign and autonomous; a *Person* above all merely human persons, in no way an object, above all circumstances and contingencies, entirely in control of everything: a *personal* God, the "Unmoved Mover." The four Levels of Being are thus seen as pointing to the invisible existence of a Level (or Levels) of Being above the human.

An interesting and instructive aspect of the progression from passivity to activity is the change in the origin of movement. It is clear that at the level of inanimate matter there cannot be movement without a physical cause, and that there is a very close linkage between cause and effect. At the plant level, the causal chain is more complex: physical causes will have physical effects as at the lower level—the wind will shake the tree whether it is living or dead—but certain physical factors act not simply as physical cause but simultaneously as *stimulus*. The sun's rays cause the plant to turn toward the sun. Its leaning too far in one direction causes the roots on the opposite side to grow stronger.

At the animal level, again, causation of movement becomes still more complex. An animal can be pushed around like a stone; it can also be stimulated like a plant; but there is, in addition, a third causative factor, which comes from inside: certain drives, attractions or compulsions, of a totally nonphysical kind—they can be called *motives*. A dog is motivated, and therefore moved, not merely by physical forces or stimuli impinging upon it from the outside, but also by forces originating in its "inner space": recognizing its master, it jumps for joy; recognizing its enemy, it runs in fear.

While at the animal level the motivating cause has to be physically present to be effective, at the level of man there is no such need. The power of self-awareness gives him an additional motivation for movement: *will*, that is, the power to move and act even when there is no physical compulsion, no physical stimulus, and no motivating force actually present. There is a lot of controversy about will: How free is it? We shall deal with this matter later. In the present context it is merely necessary to recognize that at the human level there exists an additional possibility of the origin of movement—one that does not seem to exist at any lower level, namely, movement on the basis of what might be called "naked insight." A person may move to another place not because present conditions motivate him to do so, but because he *anticipates in his mind* certain future developments.

While these additional possibilities—namely, the power of foreknowledge and therewith the capacity to anticipate future possibilities—are no doubt possessed, to some degree, by all human beings—it is evident that they vary greatly from individ-

ual to indivdual, and with most of us are very weak. It is possible
to imagine a suprahuman Level of Being where they would
exist in perfection. Perfect foreknowledge of the future would
therefore be considered a Divine attribute, associated with perfect freedom of movement and perfect freedom from passivity.
The *progression* from *physical cause* to *stimulus* to *motive* to
will would then be completed by a perfection of will capable
of overriding all the causative forces which operate at the four
Levels of Being known to us.

II

The progression from passivity to activity is similar and
closely related to the *progression* from necessity to freedom. It
is easy to see that at the mineral level there is nothing but
necessity. Inanimate matter cannot be other than what it is; it
has no choice, no possibility of "developing" or in any way
changing its nature. The so-called indeterminacy at the level of
nuclear particles is simply another manifestation of necessity,
because total necessity means the absence of any creative principle. As I have said elsewhere, it is analogous to the zero dimension—a kind of nothingness where nothing remains to be
determined. The "freedom" of indeterminacy is, in fact, the
extreme opposite of freedom: a kind of necessity which can be
understood only in terms of statistical probability. At the level
of inanimate matter, there is no "inner space" where any autonomous powers could be marshaled. As we shall see, *"inner
space" is the scene of freedom.*

We know little, if anything, about the "inner space" of plants,
more about that of animals, and a great deal about the "inner
space" of the human being: the space of a *person*, of creativity,
of *freedom*. Inner space is created by the powers of life, consciousness, and self-awareness; but we have direct and personal
experience only of our own "inner space" and the freedom it
affords *us*. Close observation discloses that most of us, most of
the time, behave and act mechanically, like machines. The spe-

cifically human power of self-awareness is asleep, and the human being, like an animal, acts—more or less intelligently—solely in response to various influences. Only when a man makes use of his power of self-awareness does he attain to the level of a person, to the level of freedom. At that moment he is living, not being lived. Numerous forces of necessity, accumulated in the past, are still determining his actions, but a small dent is being made, a tiny change of direction is being introduced. It may be virtually unnoticeable, but many moments of self-awareness can produce many such changes and even turn a given movement into the opposite of its previous direction.

To ask whether the human being *has* freedom is like asking whether man *is* a millionaire. He is not, but can become, a millionaire. He can make it his aim to become rich; similarly, he can make it his aim to become free. In his "inner space" he can develop a center of strength so that the power of his freedom exceeds that of his necessity. It is possible to imagine a perfect Being who is always and invariably exercising Its power of self-awareness, which is the power of freedom, to the fullest degree, unmoved by any necessity. This would be a Divine Being, an almighty and sovereign power, a perfect Unity.

III

There is also a marked and unmistakable *progression* toward integration and unity. At the mineral level, there is no integration. Inanimate matter can be divided and subdivided without loss of character or gestalt, simply because at this level there is none to lose. Even at plant level inner unity is so weak that parts of a plant can often be cut off, yet continue to live and develop as separate beings. Animals, by contrast, are much more highly integrated beings. Seen as a biological system, the higher animal is a unity, and parts of it cannot survive separation. There is, however, but little integration on the mental plane; that is to say, even the highest animal attains only a very modest level of logicality and consistency; its memory, on the whole, is weak, and its intellect shadowy.

Man has obviously much more inner unity than any being below him, although integration, as modern psychology recognizes, is not guaranteed to him at birth and attaining it remains one of his major tasks. As a biological system, he is most harmoniously integrated; on the mental plane, integration is less perfect but capable of considerable improvement through schooling. As a *person*, however, as a being with the power of self-awareness, he is generally so poorly integrated that he experiences himself as an assembly of many different personalities, each saying "I." The classic expression of this experience is found in Saint Paul's letter to the Romans:

> My own behaviour baffles me. For I find myself not doing what I really want to do but doing what I really loathe. Yet surely if I do things that I really don't want to do, it cannot be said that "I" am doing them at all,—it must be sin that has made its home in my nature.[1]

Integration means the creation of an inner unity, a center of strength and freedom, so that the being ceases to be a mere object, acted upon by outside forces, and becomes a subject, acting from its own "inner space" into the space outside itself. One of the greatest Scholastic statements on this *progression of integration* is found in the *Summa contra Gentiles* by Saint Thomas Aquinas:

> Of all things the inanimate obtain the lowest place, and from them no emanation is possible except by the action of one on another: thus, fire is engendered from fire when an extraneous body is transformed by fire, and receives the quality and form of fire.
>
> The next place to inanimate bodies belongs to plants, whence emanation proceeds from within, for as much as the plant's intrinsic humour is converted into seed, which being committed to the soil grows into a plant. Accordingly, here we find the first traces of life: since living things are those which move themselves to act, whereas those which can only move extraneous things are wholly lifeless. It is a sign of life in plants that something within them is the cause of

a form. Yet the plant's life is imperfect because, although in it emanation proceeds from within, that which emanates comes forth by little and little, and in the end becomes altogether extraneous: thus the humour of a tree gradually comes forth from the tree and eventually becomes a blossom, and then takes the form of fruit, distinct from the branch, though united thereto; and when the fruit is perfect it is altogether severed from the tree, and falling to the ground, produces by its seminal force another plant. Indeed if we consider the matter carefully we shall see that the first principle of this emanation is something extraneous: since the intrinsic humour of the tree is drawn through the roots from the soil whence the plant derives its nourishment.

There is yet above that of the plants a higher form of life, which is that of the sensitive soul, the proper emanation whereof, though beginning from without, terminates within. Also, the further the emanation proceeds, the more does it penetrate within: for the sensible object impresses a form on the external senses, whence it proceeds to the imagination and, further still, to the storehouse of the memory. Yet in every process of this kind of emanation, the beginning and the end are in different subjects: for no sensitive power reflects on itself. Wherefore this degree of life transcends that of plants in so much as it is more intimate; and yet it is not a perfect life, since the emanation is always from one thing to another. Wherefore the highest degree of life is that which is according to the intellect: for the intellect reflects on itself, and can understand itself. There are, however, various degrees in the intellectual life: because the human mind, though able to know itself, takes its first step to knowledge from without: for it cannot understand apart from phantasms. . . . Accordingly, intellectual life is more perfect in the angels whose intellect does not proceed from something extrinsic to acquire self-knowledge, but knows itself by itself. Yet their life does not reach the highest degree of perfection . . . because in them to understand and to be are not the same thing. . . . Therefore, the highest perfection of life belongs to God, whose understanding is not distinct from His being.[2]

This statement, unfamiliar as its mode of reasoning may be to the modern reader, makes it very clear that "higher" always means and implies "more inner," "more interior," "deeper," "more intimate"; while "lower" means and implies "more outer," "more external," "shallower," less intimate.

The more "interior" a thing is, the less visible it is likely to be. The *progression* from visibility to invisibility is just another facet of the great hierarchy of Levels of Being. There is no need to dwell on it at length. Obviously the terms "visibility" and "invisibility" refer not merely to the visual sense but to all senses of external observation. The powers of life, consciousness, and self-awareness which come into focus as we review the four Levels of Being are all wholly "invisible"—without color, sound, "skin," taste, or smell, and also without extension or weight. Nevertheless, who would deny that *they* are what we are mainly interested in? When I buy a packet of seed, my main interest is that it should be alive and not dead, and an unconscious cat, even though still alive, is not a real cat for me until it has regained consciousness. The "invisibility of man" has been incisively decribed by Maurice Nicoll:

> We can all see another person's body directly. We see the lips moving, the eyes opening and shutting, the lines of the mouth and face changing, and the body expressing itself as a whole in action. The person *himself* is invisible. . . .
>
> If the invisible side of people were discerned as easily as the visible side, we would live in a *new humanity*. As we are, we live in visible humanity, a humanity of *appearances*. . . .
>
> All our thoughts, emotions, feelings, imaginations, reveries, dreams, fantasies, are *invisible*. All that belongs to our scheming, planning, secrets, ambitions, all our hopes, fears, doubts, perplexities, all our affections, speculations, ponderings, vacuities, uncertainties, all our desires, longings, appetites, sensations, our likes, dislikes, aversions, attractions, loves and hates—all are themselves invisible. They constitute "oneself."[3]

Nicoll insists that while all this may appear obvious, it is not at all obvious: "It is an extremely difficult thing to grasp. . . . We do not grasp that we are invisible. We do not realise that we are in a world of invisible people. We do not understand *that life, before all other definitions of it, is a drama of the visible and the invisible.*"[4] There is the external world, in which things are visible, i.e., directly accessible to our senses; and there is "inner space," where things are invisible, i.e., *not directly accessible to us,* except in the case of ourselves. This all-important point will occupy us at some length in a later chapter.

The *progression* from the totally visible mineral to the largely invisible *person* can be taken as a pointer toward Levels of Being above man totally invisible to our senses. We need not be surprised that most people throughout most of human history implicitly believed in the reality of this projection; they have always claimed that just as we can learn to "see" into the invisiblity of the persons around us, so we can develop abilities to "see" the totally invisible beings existing at levels above us.

(As a philosophical mapmaker I have the duty to put these important matters on my map, so that it can be seen where they belong and how they connect with other, more familiar things. Whether or not any reader, traveler, or pilgrim wishes to explore them is his own affair.)

IV

The degree of integration, of inner coherence and strength, is closely related to the kind of "world" that exists for beings at different levels. Inanimate matter has no "world." Its total passivity is equivalent to the total emptiness of its world. A plant has a "world" of its own—a bit of soil, water, air, light, and possibly other influences—a "world" limited to its modest biological needs. The world of any one of the higher animals is incomparably greater and richer, although still mainly determined by biological needs, as modern animal psychology studies have amply demonstrated. But it also contains something more—such as curiosity, which enlarges the animal's world beyond its narrow biological confines.

The world of man, again, is incomparably greater and richer; indeed, it is asserted in traditional philosophy that man is *capax universi,* capable of bringing the whole universe into his experience. What he will actually grasp depends on each person's own Level of Being. The "higher" the person, the greater and richer is his or her world. A person, for instance, entirely fixed in the philosophy of materialistic Scientism, denying the reality of "invisibles" and confining his attention solely to what can be counted, measured, and weighed, lives in a very poor world, so poor that he will experience it as a meaningless wasteland unfit for human habitation. Equally, if he sees it as nothing but an accidental collocation of atoms, he must needs agree with Bertrand Russell that the only rational attitude is one of "unyielding despair."

It has been said: "Your Level of Being attracts your life."[5] There are no occult or unscientific assumptions behind this saying. At a low Level of Being only a very poor world exists and only a very impoverished kind of life can be lived. The Universe is what it is; but he who, although *capax universi*, limits himself to its lowest sides—to his biological needs, his creature comforts, or his accidental encounters—will inevitably "attract" a miserable life. If he can recognize nothing but "struggle for survival" and "will to power" fortified by cunning, his "world" will be one fitting Hobbes's description of the life of man as "solitary, poor, nasty, brutish, and short."

The higher the Level of Being, the greater, richer, and more wonderful is the world. If we again extrapolate beyond the human level, we can understand why the Divine was considered not merely *capax universi* but actually in total possession of it, aware of everything, omniscient: "Are not five sparrows sold for two farthings, and not one of them is forgotten before God?"[6]

If we take the "fourth dimension"—time—into consideration, a similar picture emerges. At the lowest level, there is time only in the sense of duration. For creatures endowed with consciousness there is time in the sense of experience; but experience is confined to the present, except where the past is made present through memory (of one kind or another) and the future is made present through foresight (of which, again, there may be different kinds). The higher the Level of Being, the broader, as it were, is the present, the more it embraces of what, at lower Levels of Being, is past and future. At the highest

imaginable Level of Being, there would be the *"eternal now.."*
Something like that may be the meaning of this passage in
Revelation:

> And the angel which I saw stand upon the sea and upon the earth
> lifted up his hand to heaven, and sware by him that liveth for ever
> and ever, who created heaven, and the things that therein are, and
> the earth, and the things that therein are, and the sea, and the things
> which are therein, that there should be time no longer.[7]

V

An almost infinite number of further "progressions" could be
added to those already described, but that is not the purpose of
this book. The reader will be able to fill in whatever seems to
him to be of special interest. Maybe he is interested in the
question of "final causes." Is it legitimate to explain or even to
describe a given phenomenon in teleological terms, i.e., as pur-
suing a purpose? It is ridiculous to answer such a question without reference to the Level of Being on which the phenomenon
occurs. To deny teleological action at the human level would be
as foolish as to impute it at the level of inanimate matter. Hence
there is no reason to assume that traces or remnants of teleological action may not be found at the levels in between.

The four great Levels of Being can be likened to an inverted
pyramid where each higher level comprises everything lower
and is open to influences from everything higher. All four levels
exist in the human being, which, as we have already seen, can
be described by the formula

$$\text{Man} = m + x + y + z$$
$$= \text{mineral} + \text{life} + \text{consciousness} + \text{self-awareness}$$

Not surprisingly, many teachings describe man as possessing
four "bodies," namely,

> the physical body (corresponding to m)
> the etheric body   (corresponding to x)

the astral body   (corresponding to y) and
the "I" or Ego
or Self or Spirit   (corresponding to z)

In the light of our understanding of the four great Levels of Being, such descriptions of man as a fourfold being become easily comprehensible. In some teachings, $m + x$ is taken as one —the living body (because an inanimate body is of no interest at all)—and they therefore speak of man as a threefold being, consisting of body $(m + x)$, soul (y), and Spirit (z). As people turned their interests increasingly to the visible world, the distinction between soul and Spirit became more difficult to maintain and tended to be dropped altogether; man, therefore, was represented as a being compounded of body and soul. With the rise of materialistic Scientism, finally, even the soul disappeared from the description of man—how could it exist when it could be neither weighed nor measured?—except as one of the many strange attributes of complex arrangements of atoms and molecules. Why not accept the so-called "soul"—a bundle of surprising powers—as an *epiphenomenon* of matter, just as, say, magnetism has been accepted as such? The Universe was no longer seen as a great hierarchic structure or Chain of Being; it was seen simply as an accidental collocation of atoms; and man, traditionally understood as the microcosm reflecting the macrocosm (i.e., the structure of the Universe), was no longer seen as a *cosmos*, a meaningful even though mysterious creation.

If the great Cosmos is seen as nothing but a chaos of particles without purpose or meaning, so man must be seen as nothing but a chaos of particles without purpose and meaning—a sensitive chaos perhaps, capable of suffering pain, anguish, and despair, but a chaos all the same (whether he likes it or not)—a rather unfortunate cosmic accident of no consequence whatsoever.

This is the picture presented by modern materialistic Scientism, and the only question is: Does it make sense of what we can actually experience? This is a question everybody has to decide for himself. Those who stand in awe and admiration, in wonder and perplexity, contemplating the four great Levels of Being, will not be easily persuaded that there is only *more or less*—i.e., horizontal extension. They will find it impossible to close their minds to *higher or lower*—that is to say, vertical scales and even discontinuities. If they then see man as *higher* than any arrangement, no matter how complex, of inanimate matter, and *higher* than the animals, no matter how far advanced, they will also see man as "open-ended," not at the *highest* level but with a potential that might indeed lead to perfection.

This is the most important insight that follows from the contemplation of the four great Levels of Being: At the level of man, there is no discernible limit or ceiling. Self-awareness, which constitutes the difference between animal and man, is a power of unlimited potential, a power which not only makes man human but gives him the possibility, even the need, to become superhuman. As the Scholastics used to say: *"Homo non proprie humanus sed superhumanus est"*—which means that to be properly human, you must go beyond the merely human.

4

"Adaequatio": I

What enables man to know anything at all about the world around him? "Knowing demands the organ fitted to the object," said Plotinus (died A.D. 270). Nothing can be known without there being an appropriate "instrument" in the makeup of the knower. This is the Great Truth of *"adaequatio"* (adequateness), which defines knowledge as *adaequatio rei et intellectus* —the understanding of the knower must be *adequate* to the thing to be known.

From Plotinus, again, comes: "Never did eye see the sun unless it had first become sunlike, and never can the soul have vision of the First Beauty unless itself be beautiful." John Smith the Platonist (1618–1652) said: "That which enables us to know and understand aright in the things of God, must be a living principle of holiness within us"; to which we might add the statement by Saint Thomas Aquinas that "Knowledge comes about insofar as the object known is within the knower."

We have seen already that man, in a sense, *comprises* the four great Levels of Being; there is therefore some degree of correspondence or "connaturality" between the structure of man and the structure of the world. This is a very ancient idea and has usually been expressed by calling man a "microcosm" which somehow "corresponds" with the "macrocosm" which is the world. He is a physicochemical system, like the rest of the world, and he also possesses the invisible and mysterious powers of life, consciousness, and self-awareness, some or all of which he can detect in many beings around him.

Our five bodily senses make us *adequate* to the lowest Level of Being—inanimate matter. But they can supply nothing more than masses of sense data, to "make sense" of which we require abilities or capabilities of a different order. We may call them "intellectual senses." Without them we should be unable to recognize form, pattern, regularity, harmony, rhythm, and meaning, not to mention life, consciousness, and self-awareness. While the bodily senses may be described as relatively passive, mere receivers of whatever happens to come along and to a large extent controlled by the mind, the intellectual senses are the *mind-in-action,* and their keenness and reach are qualities of the mind itself. As regards the bodily senses, all healthy people possess a very similar endowment, but no one could possibly overlook the fact that there are significant differences in the power and reach of people's minds.

It is therefore quite unrealistic to try to define and delimit the intellectual capabilities of "man" as such—as if all human beings were much the same, like animals of the same species. Beethoven's musical abilities, even in deafness, were incomparably greater than mine, and the difference did not lie in the sense of hearing; it lay in the mind. Some people are incapable of grasping and appreciating a given piece of music, not because they are deaf but because of a lack of *adaequatio* in the mind. The music is grasped by intellectual powers which some people possess to such a degree that they can grasp, and retain in their memory, an entire symphony on one hearing or one reading of the score; while others are so weakly endowed that they cannot get it at all, no matter how often and how attentively they listen to it. For the former, the symphony is as *real* as it was to the composer; for the latter, there is no symphony: there is nothing but a succession of more or less agreeable but altogether meaningless noises. The former's mind is *adequate* to the symphony; the latter's mind is *inadequate*, and thus *incapable of recognizing the existence of the symphony*.

The same applies throughout the whole range of possible and actual human experiences. For every one of us only those facts and phenomena "exist" for which we possess *adaequatio*, and as we are not entitled to assume that we are necessarily adequate to everything, at all times, and in whatever condition we may find ourselves, so we are not entitled to insist that something inaccessible to us has no existence at all and is nothing but a phantom of other people's imaginations.

There are physical facts which the bodily senses pick up, but there are also nonphysical facts which remain unnoticed *unless* the work of the senses is controlled and completed by certain "higher" faculties of the mind. Some of these nonphysical facts represent "grades of significance," to use a term coined by G. N. M. Tyrrell, who gives the following illustration:

> Take a book, for example. To an animal a book is merely a coloured shape. Any higher significance a book may hold lies above the level of its thought. And the book *is* a coloured shape; the animal is not wrong. To go a step higher, an uneducated savage may regard a book as a series of marks on paper. This is the book as seen on a higher level of significance than the animal's, and one which corresponds to the savage's level of thought. Again it is not wrong, only the book *can* mean more. It may mean a series of letters arranged according to certain rules. This is the book on a higher level of significance than the savage's. . . . Or finally, on a still higher level, the book may be an expression of meaning. . . .'

In all these cases the "sense data" are the same; the facts given to the eye are identical. Not the eye, only the mind, can determine the "grade of significance." People say: "Let the facts speak for themselves"; they forget that the speech of facts is real only if it is heard and understood. It is thought to be an easy matter to distinguish between fact and theory, between perception and interpretation. In truth, it is extremely difficult. You see the full moon just above the horizon behind the silhouettes of some trees or buildings, and it appears to you as a disc as large as that of the sun; but the full moon straight above your head looks quite small. What are the true sizes of the moon images actually received by the eye? They are exactly the same in both cases. And yet, even when you know this to be so, your mind will not easily let you see the two discs as of equal size. "Perception is not determined simply by the stimulus pattern," writes R. L. Gregory in *Eye and Brain;* "rather it is a dynamic searching for the best interpretation of the available data."[2] This searching uses not only the sensory information but also *other knowledge and experience,* although just how far experience affects perception, according to Gregory, is a difficult question to answer. In short, we "see" not simply with our eyes but with a great part of our mental equipment as well, and since this mental equipment varies greatly from person to person, there are inevitably many things which some people can "see" but which others cannot, or, to put it differently, for which some people are *adequate* while others are not.

When the level of the knower is not adequate to the level (or grade of significance) of the object of knowledge, the result is not factual error but something much more serious: an inadequate and impoverished view of reality. Tyrrell pursues his illustration further, as follows:

A book, we will suppose, has fallen into the hands of intelligent beings who know nothing of what writing and printing mean, but they are accustomed to dealing with the external relationships of things. They try to find out the "laws" of the book, which for them mean the principles governing the order in which the letters are arranged. . . . They will think they have discovered the laws of the book when they have formulated certain rules governing the external relationships of the letters. That each word and each sentence expresses a meaning will never dawn on them because their background of thought is made up of concepts which deal only with external relationships, and explanation to them means solving the puzzle of these external relationships. . . . Their methods will never reach the grade [of significance] which contains the idea of meanings.[3]

Just as the world is a hierarchic structure with regard to which it is meaningful to speak of "higher" and "lower," so the senses, organs, powers, and other "instruments" by which the human being perceives and gains knowledge of the world form a hierarchic structure of "higher" and "lower." "As above, so below," the Ancients used to say: to the world outside us there corresponds, in some fashion, a world inside us. And just as the higher levels in the world are rarer, more exceptional, than the lower levels—mineral matter is ubiquitous; life only a thin film on the Earth; consciousness, relatively rare; and self-awareness, the great exception—so it is with the abilities of people. The lowest abilities, such as seeing and counting, belong to every normal person, while the higher abilities, such as those needed for the perceiving and grasping of the more subtle aspects of reality, are less and less common as we move up the scale.

There are inequalities in the human endowment, but they are probably of much less importance than are differences in interests and in what Tyrrell calls the "background of thought." The intelligent beings of Tyrrell's allegory lacked *adaequatio* with regard to the book because they based themselves on the assumption that the "external relationships of the letters" were all that mattered. They were what we should call scientific materialists, whose faith is that objective reality is limited to that which can be actually observed and who are ruled by a *methodical aversion to the recognition of higher levels or grades of significance.*

The level of significance to which an observer or investigator tries to attune himself is chosen, not by his intelligence, but by his faith. The facts themselves which he observes do not carry labels indicating the appropriate level at which they *ought to be* considered. Nor does the choice of an inadequate level lead the intelligence into factual error or logical contradiction. All levels of significance *up to* the adequate level—i.e., up to the level of *meaning* in the example of the book—are equally factual, equally logical, equally objective, but not equally *real.*

It is by an act of faith that I choose the level of my investigation; hence the saying *"Credo ut intelligam"*—I have faith so as to be able to understand. If I lack faith, and consequently choose an inadequate level of significance for my investigation, no degree of "objectivity" will save me from missing the point of the whole operation, and I rob myself of the very possibility of understanding. I shall then be one of those of whom it has been said: "They, seeing, see not; and hearing they hear not, neither do they understand."[4]

In short, when dealing with something representing a higher grade of significance or Level of Being than inanimate matter, the observer depends not only on the adequateness of his own higher qualities, perhaps "developed" through learning and training; he depends also on the adequateness of his "faith" or, to put it more conventionally, of his fundamental presuppositions and basic assumptions. In this respect he tends to be very much a child of his time and of the civilization in which he has spent his formative years; for the human mind, generally speaking, does not just think: it thinks with ideas, most of which it simply adopts and takes over from its surrounding society.

There is nothing more difficult than to become critically aware of the presuppositions of one's thought. Everything can be seen directly except the eye through which we see. Every thought can be scrutinized directly except the thought by which we scrutinize. A special effort, an effort of self-awareness, is needed: that almost impossible feat of thought recoiling upon itself—almost impossible but not quite. In fact, this is the power that makes man human and also capable of transcending his humanity. It lies in what the Bible calls man's "inward parts." As already mentioned, "inward" corresponds with "higher" and "outward" corresponds with "lower." The senses are man's most outward instruments; when it is a case of "they, seeing, see not; and hearing they hear not," the fault lies not with the senses but with the inward parts—"for this people's *heart* is waxed gross"; they fail to *"understand with their heart."*[5] Only through the "heart" can contact be made with the higher grades of significance and Levels of Being.

For anyone wedded to the materialistic Scientism of the modern age it will be impossible to understand what this means. He has no belief in anything higher than man, and he sees in man nothing but a highly evolved animal. He insists that truth can be discovered only by means of the brain, which is situated in the head and not in the heart. All this means that "understanding with one's heart" is to him a meaningless collection of words. From his point of view, he is quite right: The brain, situated in the head and supplied with data by the bodily senses, is fully adequate for dealing with inanimate matter, the lowest of the four great Levels of Being. Indeed, its working would be only disturbed, and possibly distorted, if the "heart" interfered in any way. As a materialistic scientist, he believes that life, consciousness, and self-awareness are nothing but manifestations of complex arrangements of inanimate particles—a "faith" which makes it perfectly rational for him to place exclusive reliance on the bodily senses, to "stay in the head," and to reject any interference from the "powers" situated in the heart. For him, in other words, higher levels of Reality simply do not exist, *because his faith excludes the possibility of their existence.* He is like a man who, although in possession of a radio receiver, refuses to use it because he has made up his mind that nothing can be obtained from it but atmospheric noises.

Faith is not in conflict with reason, nor is it a substitute for reason. Faith chooses the grade of significance or Level of Being at which the search for knowledge and understanding is to aim. There is reasonable faith and there is unreasonable faith. To look for meaning and purpose at the level of inanimate matter would be as unreasonable an act of faith as an attempt to "explain" the masterpieces of human genius as nothing but the outcome of economic interests or sexual frustrations. The faith of the agnostic is perhaps the most unreasonable of all, because, unless it is mere camouflage, it is a decision to treat the question of significance as insignificant, like saying: "I am not willing to decide whether [reverting to Tyrrell's example] a book is merely a colored shape, a series of marks on paper, a series of letters arranged according to certain rules, or an expression of meaning." Not surprisingly, traditional wisdom has always treated the agnostic with withering contempt: "I know thy works, that thou art neither cold nor hot: I would thou wert cold or hot. So then because thou art lukewarm, and neither cold nor hot, I will spue thee out of my mouth."[6]

It can hardly be taken as an unreasonable act of faith when people accept the testimony of prophets, sages, and saints who, in different languages but with virtually one voice, declare that the book of this world is not merely a colored shape but an expression of meaning; that there are Levels of Being above that of humanity; and that man can reach these higher levels provided he allows his reason to be guided by faith. No one has described man's possible journey to the truth more clearly than the Bishop of Hippo, Saint Augustine (354–430):

> The first step forward . . . will be to see that the attention is fastened on truth. Of course faith does not see truth clearly, but it has an eye for it, so to speak, which enables it to see that a thing is true even when it does not see the reason for it. It does not yet see the thing it believes, but at least it knows for certain that it does not see it and that it is true none the less. This possession through faith of a hidden but certain truth is the very thing which will impel the mind to penetrate its content, and to give the formula, "Believe that you may understand" *(Crede ut intelligas)*, its full meaning.[7]

With the light of the intellect we can see things which are invisible to our bodily senses. No one denies that mathematical and geometrical truths are "seen" in this way. To *prove* a proposition means to give it a form, by analysis, simplification, transformation, or dissection, through which the truth can be *seen;* beyond this seeing there is neither the possibility of nor the need for any further proof.

Can we see, with the light of the intellect, things which go beyond mathematics and geometry? Again, no one denies that we can *see* what another person means, sometimes even when he does not express himself accurately. Our everyday language is a constant witness to this power of *seeing,* of grasping ideas, which is quite different from the processes of thinking and forming opinions. It produces flashes of understanding.

> As far as St Augustine is concerned, faith is the heart of the matter. *Faith tells us what there is to understand;* it purifies the heart, and so allows reason to profit from discussion; it enables reason to arrive at an understanding of God's revelation. In short, when Augustine

speaks of understanding, he always has in mind the product of a rational activity for which faith prepares the way.[8]

As the Buddhists say, faith opens "the eye of truth," also called "the Eye of the Heart" or "the Eye of the Soul." Saint Augustine insisted that "our whole business in this life is to restore to health the eye of the heart whereby God may be seen." Persia's greatest Sufi poet, Rumi (1207–1273), speaks of "the eye of the heart, which is seventy-fold and of which these two sensible eyes are only the gleaners";[9] while John Smith the Platonist advises: "We must shut the eyes of sense, and open that brighter eye of our understandings, that other eye of the soul, as the philosopher calls our intellectual faculty, 'which indeed all have, but few make use of it.' "[10] The Scottish theologian, Richard of Saint-Victor (d. 1173), says: "For the outer sense alone perceives visible things and the eye of the heart alone sees the invisible."[11]

The power of "the Eye of the Heart," which produces *insight,* is vastly superior to the power of thought, which produces *opinions.* "Recognising the poverty of philosophical opinions," says the Buddha, "not adhering to any of them, seeking the truth, *I saw.* "[12] The process of mobilizing the various powers possessed by man, gradually and, as it were, organically, is described in a Buddhist text:

> One can not, I say, attain supreme knowledge all at once; only by a gradual training, a gradual action, a gradual unfolding, does one attain perfect knowledge. In what manner? A man comes, moved by confidence; having come, he joins; having joined, he listens; listening, he receives the doctrine; having received the doctrine, he remembers it; he examines the sense of the things remembered; from examining the sense, the things are approved of; having approved, desire is born; he ponders; pondering, he eagerly trains himself; and eagerly training himself, he mentally realises the highest truth itself and, penetrating it by means of wisdom, *he sees.*[13]

This is the process of gaining *adaequatio,* of developing the instrument capable of seeing and thus understanding the truth that does not merely inform the mind but liberates the soul.

"And ye shall know the truth, and the truth shall make you free."[14]

As these matters have become unfamiliar in the modern world, it may be of value if I quote a contemporary author, Maurice Nicoll:

> A world of *inward* perception then begins to open out, distinct from that of outer perception. Inner space appears. *The creation of the world begins in man himself.* At first all is darkness: then light appears and is separated from the darkness. By this light we understand a form of consciousness to which our ordinary consciousness is, by comparison, darkness. This light has constantly been equated with truth and freedom. Inner perception of oneself, of one's invisibility, is the beginning of light. This perception of truth is not a matter of sense-perception, but of the perception of the truth of "ideas"—through which, certainly, the perception of our senses is greatly increased. The path of self-knowledge has this aim in view, for no one can know himself unless he turns inwards. . . . This struggle marks the commencement of that inner development of man which has been written about in many different ways (yet really always in the same way) throughout that small part of Time whose literature belongs to us, and which we think of as the entire history of the world.[15]

We shall take a closer look at the process of "turning inward" in a later chapter. Here it is necessary simply to recognize that sense data alone do not produce insight or understanding of any kind. *Ideas* produce insight and understanding, and the world of ideas lies within us. The truth of ideas cannot be seen by the senses but only by that special instrument sometimes referred to as "the Eye of the Heart," which, in a mysterious way, has the power of recognizing truth when confronted with it. If we describe the results of this power as illumination, and the results of the senses as experience, we can say that

1. Experience, and not illumination, tells us about the existence, appearance, and changes of sensible things, such as stones, plants, animals, and people.
2. Illumination, and not experience, tells us what such things

mean, what they could be, and what they perhaps ought to be.

Our bodily senses, yielding experience, do not put us into touch with the higher grades of significance and the higher Levels of Being existing in the world around us: they are not *adequate* for such a purpose, having been designed solely for registering the *outer* differences between various existing things and not their *inner* meanings.

There is a story of two monks who were passionate smokers and who tried to settle between themselves the question of whether it was permissible to smoke while praying. As they could come to no conclusion, they decided to ask their respective superiors. One of them got into deep trouble with his abbot; the other received a pat of encouragement. When they met again, the first one, slightly suspicious, inquired of the second: "What did you actually ask?" and received the answer "I asked whether it was permissible to pray while smoking." While our inner senses infallibly see the profound difference between "praying while smoking" and "smoking while praying," to our outer senses there is no difference at all.

Higher grades of significance and Levels of Being cannot be recognized without faith and the help of the higher abilities of the inner man. When these higher abilities are not brought into action, either because they are lacking or because an absence of faith leaves them unutilized, there is a lack of *adaequatio* on the part of the knower, with the consequence that nothing of higher significance or Level of Being can be known by him.

5

"Adaequatio": II

The Great Truth of *"adaequatio"* affirms that nothing can be perceived without an appropriate organ of perception and that nothing can be understood without an appropriate organ of understanding. For cognition at the mineral level, man's primary instruments are his five senses, reinforced and extended by a great array of ingenious apparatus. They register the visible world, but cannot register the "inwardness" of things and such fundamental invisible powers as life, consciousness, and self-awareness. Who could see, hear, touch, taste, or smell *life as such?* It has no shape or color, no specific sound or texture or taste or smell. And yet as we *are* able to recognize life, we must have an organ of perception to do so, an organ more inward—and that means "higher"—than the senses. We shall see later that this "organ" is the life inside ourselves, the unconscious vegetative processes and feelings of our living body, centered mainly in the solar plexus. Similarly, we recognize *consciousness* directly with our own consciousness, centered mainly in the head; and we recognize *self-awareness* with our own self-awareness, which resides, in a sense that is both symbolical and also verifiable by physical experience, in the heart region, the innermost and therefore "highest" center of the human being.

The answer to the question "What are man's instruments by which he knows the world outside him?" is therefore quite inescapably this: "Everything he has got"—his living body, his mind, and his self-aware Spirit.

Since Descartes we have been inclined to believe that we know even of our existence only through our head-centered thinking—*"Cogito ergo sum"*—I think and thus I know I exist. But every craftsman realizes that his power of knowing consists not only of the thinking in his head but also of the intelligence of his body: his fingertips know things that his thinking knows nothing about, just as Pascal knew that "The heart has its reasons which reason knows nothing about." It may even be misleading to say that man has many instruments of cognition, since, in fact, the *whole man* is one instrument. If he persuades himself that the only "data" worth having are those delivered by his five senses, and that a "data-processing unit" called the brain is there to deal with them, he restricts his knowing to that Level of Being for which these instruments are *adequate,* and this means mainly to the level of inanimate matter.

It was Sir Arthur Eddington (1882–1944) who said: "Ideally, all our knowledge of the universe could have been reached by visual sensation alone—in fact by the simplest form of visual sensation, colourless and non-stereoscopic."[1] If this is true (as it well may be), if the scientific picture of the Universe is the result of the use of the sense of sight only, restricted to the use of "a single, colour-blind eye," we can hardly expect that picture to show more than an abstract, inhospitable mechanism. The Great Truth of *adaequatio* teaches us that restriction in the use of instruments of cognition has the inevitable effect of narrowing and impoverishing reality. Surely, nobody *wishes* to obtain this effect. How, then, can it be explained that such a narrowing has taken place?

To answer this question, we have to turn again to the father of the modern development, Descartes. He was not a man lacking self-confidence. "The true principles," he said, "by which we can attain the highest degree of wisdom, which constitutes the sovereign good of human life, are those I have put in this book." "Man has . . . had many opinions so far; he has never had 'the certain knowledge of anything.' . . . But now he reaches manhood, he becomes master of himself and capable of adjusting everything to the level of reason." So Descartes claims to lay the foundations of "the admirable science," which is built upon those *"ideas easiest to grasp, the simplest, and which can be most directly represented."*[2] And what, in the end, is easiest to grasp, simplest, and capable of being most directly represented? Precisely the "pointer readings"* against a quantitative scale highlighted by Sir Arthur Eddington.

The sense of sight, restricted to the use of a single color-blind eye, being the lowest, most outward, and most superficial (i.e., surface-bound) of man's instruments of cognition, is available equally to every normal person, as is the ability to count. Needless to say, to *understand* the significance of data thus obtained requires some of the higher, and therefore rarer, faculties of the mind; but this is not the point. The point is that once a theory has been advanced—perhaps by a man of genius—anyone who takes the necessary trouble can "verify" it. Knowledge obtainable from "pointer readings" is therefore *"public knowledge,"* available to anyone, precise, indubitable, easy to check, easy to communicate, above all: *virtually untainted by any subjectivity on the part of the observer.*

I said earlier that it is often extremely difficult to get at bare facts unmingled with thoughts, adjustments, or adaptations preexisting in the observer's mind. But what can the mind add to pointer readings made by a single color-blind eye? What can it add to counting? Restricting ourselves to this mode of observation, we can indeed eliminate subjectivity and attain objectivity. Yet one restriction entails another: We attain objectivity, but we fail to attain knowledge of the *object as a whole*. Only

---

*Cf. Ernst Lehrs, *Man or Matter*, London, 1951. "In fact, physical science is essentially, as Professor Eddington put it, a 'pointer reading science.' Looking at this fact in our way we can say that all pointer instruments which man has constructed ever since the beginning of science, have as their model man himself, restricted to colourless, non-stereoscopic observation. For all that is left to him in this condition is to focus points in space and register changes of their positions. Indeed, the perfect scientific observer is himself the arch-pointer-instrument." (Pages 132–33.)

the "lowest," the most superficial, aspects of the object are accessible to the instruments we employ; everything that makes the object humanly interesting, meaningful and significant escapes us. Not surprisingly, the world picture resulting from this method of observation is "the abomination of desolation," a wasteland in which man is a quaint cosmic accident signifying nothing.

Descartes wrote:

> . . . it is the mathematicians alone who have been able to find demonstrations. . . . I did not doubt that I must start with the same things that they have considered. . . . The long chains of perfectly simple and easy reasons which geometers are accustomed to employ in order to arrive at their most difficult demonstrations, had given me reason to believe that all things which fall under the knowledge of man succeed each other in the same way and that . . . there can be none so remote that they may not be reached, or so hidden that they may not be discovered.[3]

It is obvious that a mathematical model of the world—which is what Descartes was dreaming about—can deal only with factors that can be expressed as interrelated quantities. It is equally obvious that (while *pure quantity* cannot exist) the quantitative factor is of preponderant weight at the lowest Level of Being. As we move up the Chain of Being, the importance of quantity recedes while that of quality increases, and the price of mathematical model-building is the loss of the qualitative factor, the very thing that matters most.

The change of Western man's interest from "the slenderest knowledge that may be obtained of the highest things" (Thomas Aquinas) to mathematically precise knowledge of lesser things —"there being nothing in the world the knowledge of which would be more desirable or more useful" (Christian Huygens, 1629–1695)—marks a shift from what we might call "science for understanding" to "science for manipulation." The purpose of the former was the enlightenment of the person and his "liberation"; the purpose of the latter is power. "Knowledge itself is power," said Francis Bacon, and Descartes promised men they would become "masters and possessors of nature." In its more sophisticated development, "science for manipulation" tends almost inevitably to advance from the manipulation of nature to that of people.

"Science for understanding" has often been called *wisdom,* while the name "science" remained reserved for what I call "science for manipulation." Saint Augustine, among many others, makes this distinction, and Etienne Gilson paraphrases him as follows:

> The real difference which sets the one against the other derives from the nature of their objects. The object of wisdom is such that, by reason of its intelligibility alone, no evil use can be made of it; the object of science is such that it is in constant danger of falling into the clutches of cupidity, owing to its very materiality. Hence the double designation we may give science according as it is subservient to appetite, as it is whenever it chooses itself as its end, or is subservient to wisdom, as it is whenever it is directed towards the sovereign good.[4]

These points are of crucial importance. When "science for manipulation" is subordinated to wisdom, i.e., "science for understanding," it is a most valuable tool, and no harm can come of it. But it cannot be so subordinated when wisdom disappears because people cease to be interested in its pursuit. This has been the history of Western thought since Descartes. The old science—"wisdom" or "science for understanding"—was directed primarily "towards the sovereign good," i.e., the True, the Good, and the Beautiful, knowledge of which would bring both happiness and salvation. The new science was mainly directed toward material power, a tendency which has meanwhile developed to such lengths that the enhancement of political and economic power is now generally taken as the first purpose of, and main justification for, expenditure on scientific work. The old science looked upon nature as God's handiwork and man's mother; the new science tends to look upon nature as an adversary to be conquered or a resource to be quarried and exploited.

The greatest and most influential difference, however, springs from science's view of man. "Science for understanding" saw man as made in the image of God, the crowning glory of creation, and hence "in charge of" the world, because *Noblesse oblige.* "Science for manipulation," inevitably, sees man as nothing but an accidental product of evolution, a higher animal, a social animal, and an object for study by the same methods by which other phenomena of this world are to be studied, "objectively." Wisdom is a type of knowledge that can be gained only by bringing into play the highest and noblest powers of the mind; "science for manipulation," by contrast, is a type of knowledge that can be gained by bringing into play only such powers as are possessed by everybody (except the severely handicapped), mainly pointer reading and counting, without any need to understand why a formula works: to know that it *does* work is enough for practical and manipulative purposes.

This type of knowledge is therefore public, i.e., describable in terms of general validity, so that, when correctly described, everybody can recognize it. Such public and "democratic" availability cannot be attained by knowledge relating to the higher Levels of Being, simply because the latter is not describable in terms to which everybody is *adequate.* It is claimed that only such knowledge can be termed "scientific" and "objective" as is open to public verification or falsification by anybody who takes the necessary trouble; all the rest is dismissed as "unscientific" and "subjective." The use of these terms in this manner is a grave abuse, for all knowledge is "subjective" inasmuch as it cannot exist otherwise than in the mind of a human subject, and the distinction between "scientific" and "unscientific" knowledge is question-begging, the only valid question about knowledge being that of its truth.

The progressive elimination of "science for understanding" —or "wisdom"—from Western civilization turns the rapid and ever-accelerating accumulation of "knowledge for manipulation" into a most serious threat. As I have said in another context, "We are now far too clever to be able to survive without wisdom," and further expansion of our cleverness can be of no benefit whatever. The steadily advancing concentration of man's scientific interest on "sciences of manipulation" has at least three very serious consequences.

First, in the absence of sustained study of such "unscientific" questions as "What is the meaning and purpose of man's existence?" and "What is good and what is evil?" and "What are man's absolute rights and duties?" a civilization will necessarily and inescapably sink ever more deeply into anguish, despair, and loss of freedom. Its people will suffer a steady decline in health and happiness, no matter how high may be their standard of living or how successful their "health service" in prolonging their lives. It is nothing more nor less than a matter of "Man cannot live by bread alone."

Second, the methodical restriction of scientific effort to the most external and material aspects of the Universe makes the world look so empty and meaningless that even those people who recognize the value and necessity of a "science of understanding" cannot resist the hypnotic power of the allegedly scientific picture presented to them and lose the courage as well as the inclination to consult, and profit from, the "wisdom tradition of mankind." Since the findings of science, on account of its methodical restriction and its systematic disregard of higher levels, never contain any evidence of the existence of such levels, the process is self-reinforcing: faith, instead of being taken as a guide leading the intellect to an understanding of the higher levels, is seen as opposing and rejecting the intellect and is therefore itself rejected. Thus all roads to recovery are barred.

Third, the higher powers of man, no longer being brought into play to produce the knowledge of wisdom, tend to atrophy and even disappear altogether. As a result, all the problems which society or individuals are called upon to tackle become insoluble. Efforts grow ever more frantic, while unsolved and seemingly insoluble problems accumulate. While wealth may continue to increase, the quality of man himself declines.

II

In the ideal case, the structure of a man's knowledge would match the structure of reality. At the highest level there would be "knowledge for understanding" in its purest form; at the lowest there would be "knowledge for manipulation." Understanding is required to decide what to do; the help of "knowledge for manipulation" is needed to act effectively in the material world.

For successful action, we need to know the probable results of alternative courses of action, so we can select the course most suitable for our purposes. At this level, therefore, it is correct to say that the goal of knowledge is prediction and control. The pursuit of science is a matter of taking stock and formulating recipes for action. Every recipe is a conditional sentence of the type "If you want to achieve this or that, take such and such steps." The sentence should be as concise as possible, containing no ideas or concepts that are not strictly necessary ("Ockham's razor"), and the instructions should be precise, leaving as little as possible to the judgment of the operator. The test of a recipe is purely pragmatic, the proof of the pudding being in the eating. The perfections of this type of science are purely practical, objective—i.e., independent of the character and interests of the operator, measurable, recordable, repeatable. Such knowledge is "public" in the sense that it can be used even by evil men for nefarious purposes, it gives power to anyone who manages to get hold of it. (Not surprisingly, therefore, attempts are always being made to keep parts of this "public" knowledge secret!)

At the higher levels, the very ideas of prediction and control become increasingly objectionable and even absurd. The theologian, who strives to obtain knowledge of Levels of Being above the human, does not for a moment think of prediction, control, or manipulation. All he seeks is understanding. He would be shocked by predictabilities. Anything predictable can be so only on account of its "fixed nature," and the higher the Level of Being, the less is the fixity and the greater the plasticity of nature. "With God all things are possible,"[5] but the freedom of action of a hydrogen atom is exceedingly limited. The sciences of inanimate matter—physics, chemistry, and astronomy—can therefore achieve virtually perfect powers of prediction; they can, in fact, be completed and finalized, once and for all, as is claimed to be the case with mechanics.

Human beings are highly predictable as physicochemical systems; less so as living bodies; much less so as conscious beings; and hardly at all as self-aware persons. The reason for this unpredictability does not lie in a lack of *adaequatio* on the part of the investigator, but in the nature of freedom. In the face of freedom, "knowledge for manipulation" is impossible, but "knowledge for understanding" is indispensable. The almost complete disappearance of the latter from Western civilization is due to nothing but the systematic neglect of traditional wisdom, of which the West has as rich a store as any other part of mankind. The result of the lopsided development of the last three hundred years is that Western man has become rich in means and poor in ends. The hierarchy of his knowledge has been decapitated: his will is paralyzed because he has lost any grounds on which to base a hierarchy of values. What are his highest values?

A man's highest values are reached when he claims that something is a good in itself, requiring no justification in terms of any higher good. Modern society prides itself on its "pluralism," which means that a large number of things are admissible as "good in themselves," as ends rather than as means to an end. They are all of equal rank, all to be accorded *first priority*. If something that requires no justification may be called an "absolute," the modern world, which *claims* that everything is relative, does, in fact, worship a very large number of "absolutes." It would be impossible to compile a complete list, and we shall not attempt it here. Not only power and wealth are treated as good in themselves—provided they are mine, and not someone else's—but also knowledge for its own sake, speed of movement, size of market, rapidity of change, quantity of education, number of hospitals, etc., etc. In truth, none of these sacred cows is a genuine end; they are all means parading as ends. "In the *Inferno* of the world of knowledge," comments Etienne Gilson,

> there is a special punishment for this sort of sin; it is a relapse into mythology. . . . A world which has lost the Christian God cannot but resemble a world which had not yet found him. Just like the world of Thales and of Plato, our modern world is "full of gods." There are blind Evolution, clear-sighted Orthogenesis, benevolent Progress, and others which it is more advisable not to name. Why unnecessarily hurt the feelings of men who, today, render them a cult? It is however important for us to realise that mankind is doomed to live more and more under the spell of a new scientific, social, and political mythology, unless we resolutely exorcise these befuddled notions whose influence on modern life is becoming appalling. . . . For when gods fight among themselves, men have to die.[6]

When there are so many gods, all competing with one another and claiming first priority, and there is no supreme god, no supreme good or value in terms of which everything else needs to justify itself, society cannot but drift into chaos. The modern world is full of people whom Gilson describes as "pseudo-agnostics who . . . combine scientific knowledge and social generosity with a complete lack of philosophical culture."[7] They ruthlessly use the prestige of "science for manipulation" to *discourage* people from trying to restore wholeness to the edifice of human knowledge by developing—*redeveloping*—a "science for understanding."

Is it fear that motivates them? Abraham Maslow suggests that the pursuit of science is often a defense. "It can be primarily a safety philosophy, a security system, a complicated way of avoiding anxiety and upsetting problems. In the extreme instance it can be a way of avoiding life, a kind of self-cloistering."[8]

However that may be, and it is not our task and purpose to study the psychology of scientists, there is undoubtedly also an urgent desire to escape from any traditional notions of human duties, responsibilities, or obligations the neglect of which may be sinful. In spite of the modern world's chaos and its suffering, there is hardly a concept more unacceptable to it than the idea of sin. What could be the meaning of sin anyhow? Traditionally, it means "missing the mark," as in archery, missing the very purpose of human life on earth, a life that affords unique opportunities for development, a great chance and privilege, as the Buddhists have it, "hard to obtain." Whether tradition speaks the truth or not cannot be decided by any "science for manipulation"; it can be decided only by those highest faculties of man which are *adequate* to the creation of a "science for understanding." If the very possibility of the latter is systematically denied, the highest faculties are never brought into play, they atrophy, and the very possibility of first understanding and then fulfilling the purpose of life disappears.

William James (1842–1910) was under no illusion on the point that, for each of us, this matter is primarily a question of our will—as indeed *faith* is seen traditionally as a matter of the will:

> The question of having moral beliefs at all or not having them, is decided by our will. Are our moral preferences true or false, or are they only odd biological phenomena, making things good or bad for *us*, but in themselves indifferent? How can your pure intellect decide? If your heart does not *want* a world of moral reality, your head will assuredly never make you believe in one. Mephistophelian scepticism, indeed, will satisfy the head's play-instincts much better than any rigorous idealism can.[9]

The modern world tends to be skeptical about everything that makes demands on man's higher faculties. But it is not at all skeptical about skepticism, which demands hardly anything.

6

The Four Fields of Knowledge: 1

The first landmark we have chosen for the construction of our philosophical map and guidebook is the hierarchic structure of the world—four great Levels of Being, in which the higher level always "comprehends" the levels below it.

The second landmark is the similar (in the sense of "corresponding") structure of human senses, abilities, and cognitive powers, for we cannot experience any part or facet of the world unless we possess and use an organ or instrument through which we are able to receive what is being offered. If we do not have the requisite organ or instrument, or fail to use it, we are not *adequate* to this particular part or facet of the world, with the result that, as far as we are concerned, it simply does not exist. This is the Great Truth of *"adaequatio."*

It follows from this truth that any systematic neglect or restriction in the use of our organs of cognition must inevitably have the effect of making the world appear less meaningful, rich, interesting, and so on than it actually is. The opposite is equally true: the use of organs of cognition which for one reason or another normally lie dormant, and their systematic development and perfection, enable us to discover new meaning, new riches, new interests—facets of the world which had previously been inaccessible to us.

We have seen that the modern sciences, in a determined effort to attain objectivity and precision, have indeed restricted the use of the human instruments of cognition in a rather extreme way: according to some scientific interpreters, to observations of quantitative scales by color-blind, non-stereoscopic vision. Such a methodology necessarily produces a picture of the world virtually confined to the lowest level of manifestation, that of inanimate matter, and tends to suggest that the higher Levels of Being, including human beings, are really nothing more than atoms in somewhat complex arrangements. We shall now pursue this matter a bit further. If the current methodology produces an incomplete, one-sided, and grossly impoverished picture, what methods need to be applied to obtain the full picture?

It has often been observed that for every one of us reality splits into two parts: Here am I; and there is everything else, the world, including you.

We have also had occasion to observe another duality: there are visibilities and invisibilities or, we might say, outer appearances and inner experiences. The latter become relatively more and more important than the former as we move up the Great Chain of Being. Though inner experiences unquestionably exist, they cannot be observed by our ordinary senses.

From these two pairs

"I" and "The World"
"Outer Appearance" and "Inner Experience"

we obtain four "combinations," which we can indicate thus:

1. I—inner   3. I—outer
2. The world (you)—inner   4. The world (you)—outer.

These are the *Four Fields of Knowledge,* each of which is of great interest and importance to every one of us. The four questions which lead to these fields of knowledge may be put like this:

1. What is really going on in my own inner world?
2. What is going on in the inner world of other beings?

3. What do I look like in the eyes of other beings?
4. What do I actually observe in the world around me?

To simplify in an extreme manner we might say:

1. What do I feel like?
2. What do you feel like?
3. What do I look like?
4. What do you look like?

(The numbering of these four questions, and consequently of the Four Fields of Knowledge is, of course, quite arbitrary.)

Now, the first point to make about these Four Fields of Knowledge is that we have direct access to only two of them —Field 1 and Field 4. That is to say, I can directly feel what I feel like, and I can directly see what you look like; but what it feels like to be you, I cannot directly know; and what I look like in your eyes, I do not know either. How we obtain knowledge of the other two fields—2 and 3—which are *not* directly accessible to us—that is, how we come to know and understand what is going on *inside* other beings (Field 2) and what we ourselves are *from the outside*, simply as an object of observation, as one being among countless other beings (Field 3) —how we enter these two fields of knowledge is indeed one of the most interesting, and also vital, questions that can be posed.

Socrates (in Plato's *Phaedrus*) says: "I must first know myself, as the Delphian inscription says; to be curious about that which is not my concern, while I am still in ignorance of my own self, would be ridiculous." Let us follow this example and start with *Field of Knowledge No. 1:* What, really, *is* going on inside myself? What gives me joy, what gives me pain? What strengthens me and what weakens me? Where do I control life and where does life control me? Am I in control of my mind, my feelings, can I do what I want to do? What is the value of this inner knowledge for the conduct of my life?

Before we go into any details we should take cognizance of the fact that the above-quoted statement from Plato's *Phaedrus* can be matched by similar statements from all parts of the world and all times. I shall confine myself to a few:

From Alexandria, Philo Judaeus (late first century B.C.):

> For pray do not . . . spin your airy fables about moon or sun or
> the other objects in the sky and in the universe so far removed
> from us and so varied in their natures, until you have scrutinised
> and come to know yourselves. After that, we may perhaps be-
> lieve you when you hold forth on other subjects; but before you
> establish who you yourselves are, do not think that you will ever
> become capable of acting as judges or trustworthy witnesses in
> the other matters.

From ancient Rome, Plotinus (A.D. 205?–270):

> Withdraw into yourself and look. And if you do not find yourself
> beautiful yet, act as does the creator of a statue that is to be made
> beautiful; he cuts away here, he smoothes there, he makes this line
> lighter, this other purer, until a lovely face has grown upon his work.
> So do you also: . . . never cease chiseling your statue.

From medieval Europe, the *Theologia Germanica* (ca. A.D.
1350):

> Thoroughly to know oneself, is above all art, for it is the highest
> art. If thou knowest thyself well, thou art better and more praise-
> worthy before God, than if thou didst not know thyself, but didst
> understand the course of the heavens and of all the planets and stars,
> also the virtue of all herbs, and the structure and dispositions of all
> mankind, also the nature of all beasts, and, in such matters, hadst all
> the skill of all who are in heaven and on earth.

Paracelsus (1493?–1541), who was one of the most knowledgeable men in the Europe of his time and foremost in knowing
"the virtue of all herbs," says:

> Men do not know themselves, and therefore they do not under-
> stand the things of their inner world. Each man has the essence of
> God and all the wisdom and power of the world (germinally) in
> himself; he possesses one kind of knowledge as much as another, and
> he who does not find that which is in him cannot truly say that he
> does not possess it, but only that he was not capable of successfully
> seeking for it.

From India, Swami Ramdas (1886–1963):

> "Seek within—know thyself," these secret and sublime hints come to us wafted from the breath of Rishis through the dust of ages.

From the world of Islam, Azid ibn Muhammad al-Nasafi (seventh-eighth centuries):

> When 'Ali asked Mohammad, "What am I to do that I may not waste my time?" the Prophet answered, "Learn to know thyself."

And from China, the *Tao Tê Ching* by Lao-tse (c. 604–531 B.C.):

> He who knows others is wise;
> He who knows himself is enlightened.[1]

Finally, let us listen to a twentieth-century writer, P. D. Ouspensky (1878–1947), who states as his "fundamental idea":

> that man as we know him *is not a completed being;* that nature develops him only up to a certain point and then leaves him, either to develop further, *by his own* efforts and devices, or to live and die such as he was born, or to degenerate and lose capacity for development.
>
> Evolution of man . . . will mean the development of certain *inner* qualities and features which usually remain undeveloped, *and cannot develop by themselves.*[2]

The modern world knows little of all this, even though it has produced more psychological theories and literature than any previous age. As Ouspensky says: "Psychology is sometimes called a new science. This is quite wrong. Psychology is, perhaps, the *oldest science,* and, unfortunately, in its most essential features a *forgotten science.*" These "most essential features" presented themselves primarily in religious teachings, and their disappearance is accounted for largely by the decline of religion during the last few centuries.

Traditional psychology, which saw people as "pilgrims" and "wayfarers" on this earth who could reach the summit of a mountain of "salvation," "enlightenment," or "liberation," was primarily concerned not with sick people who had to be made "normal" but with normal people who were capable of becoming, and indeed destined to become, supernormal. Many of the great traditions have the idea of "The Way" at their very center: the Chinese teaching of Taoism is named after *Tao*, "The Way"; the Buddha's teaching is called "The Middle Way"; and Jesus Christ Himself declares: "I am the Way." It is the pilgrim's task to undertake a *journey into the interior* which demands a degree of heroism and in any case a readiness occasionally to turn one's back on the petty preoccupations of everyday life. As Joseph Campbell shows in his wonderful study of *The Hero with a Thousand Faces*, the traditional teachings, most of which are in the form of mythology, do "not hold as [their] greatest hero the merely virtuous man. Virtue is but the pedagogical prelude to the culminating insight, which goes beyond all pairs of opposites."[3] Only a perfectly clean instrument can obtain a perfectly clear picture.

It should not be thought that the *journey into the interior* is only for heroes. It requires an inner commitment, and there is something heroic about any commitment to the unknown, but it is a heroism within everybody's capability. It is obvious that the study of this "First Field of Knowledge" demands the whole person, for only a whole person can be adequate to the task. A one-eyed, color-blind observer would certainly not get very far. But how can the whole person—which means the human being's highest qualities—be brought into play? In discussing the four Levels of Being we found that the enormous superiority of the human over the animal level needed to be acknowledged; and the "additional power"—"z"—which accounted for man's superiority over the animals, we identified as being closely connected with *self-awareness*. Without self-awareness the exploration and study of the inner man, i.e., one's interior world, is completely impossible.

Now, self-awareness is closely related to the power of attention, or perhaps I should say the power of *directing* attention. My attention is often, or perhaps most of the time, *captured* by outside forces which I may or may not have chosen myself—sights, sounds, colors, etc.—or else by forces inside myself—expectations, fears, worries, interests, etc. When it is so captured,

I function very much like a machine: I am not *doing* things; they simply *happen.* All the time, there exists, however, the possibility that I may take the matter in hand and quite freely and deliberately *direct* my attention to something entirely of my own choosing, something that does not capture me but is to be captured by me. The difference between directed and captured attention is the same as the difference between doing things and letting things take their course, or between living and "being lived." No subject could be of greater interest; no subject occupies a more central place in all traditional teachings; and no subject suffers more neglect, misunderstanding, and distortion in the thinking of the modern world.

In his book on *Yoga,* Ernest Wood talks about a state which he (wrongly, I believe) calls contemplation:

> Yes we often "lose ourselves." We peep into someone's office or study, and tip-toe away, whispering to our companions, "He is lost in thought." I knew a man who used to lecture frequently, on subjects requiring much thought. He told me that he had acquired the power to put himself out of mind—completely forget himself—at the commencement of a lecture, and look mentally at his subject-matter like a map on which he was following a route, while the spoken words flowed in complete obedience to the successive ideas which were being looked at. He told me that he would become aware of himself perhaps once or twice during the lecture, and at the end of it, as he sat down, he would find himself surprised that it was he who had given the lecture. Yet he fully remembered everything.[4]

This is a very good description of a man acting like a programmed machine, implementing a program devised some time previously. *He,* the programmer, is no longer needed; he can mentally absent himself. If the machine is implementing a good program, it gives a good lecture; if the program is bad, the lecture is bad. We are all familiar with the possibility of implementing "programs," e.g., driving a car and engaging in an interesting conversation at the same time: paradoxically, we may be driving "attentively," carefully, considerately, yet all our real attention is on the conversation. Are we equally familiar with directing our attention to where we want it to be, not depending on any "attraction," and keeping it there for as long as we desire? We are not. Such moments of full freedom and self-awareness are all too rare. Most of our life is spent in some kind of thralldom; we are captivated by this or that, drift along in our captivity, and carry out programs which have been lodged in our machine, we do not know how, when, or by whom.

The first subject for study in what I have called "Field 1" is therefore *attention*, and this leads immediately to a study of our mechanicalness. The best help in this study that I know of is P. D. Ouspensky's book on *The Psychology of Man's Possible Evolution*.

It is not difficult to verify for oneself Ouspensky's observation that we may at any time find ourselves in any one of three different "states" or "parts of ourselves"—mechanical, emotional, or intellectual. The chief criterion for identifying these different "parts" is the quality of our attention. "Without attention or with attention wandering, we are in the mechanical part; with the attention attracted by the subject of observation or reflection and kept there, we are in the emotional part; with the attention controlled and held on the subject by will, we are in the intellectual part."[5]

Now, in order to be aware of where our attention is and what it is doing, we have to be *awake* in a rather exacting meaning of the word. When we are acting or thinking or feeling *mechanically*, like a programmed computer or any other machine, we are obviously not awake in that sense, and we are doing, thinking, or feeling things which we have not ourselves freely chosen to do, think, or feel. We may say afterward: "I did not mean to do it" or "I don't know what came over me." We may intend, undertake, and even solemnly promise to do all kinds of things, but if we are at any time liable to drift into actions "we did not mean to do" or to be pushed by some thing that "comes over us," what is the value of our intentions? When we are not awake in our attention, we are certainly not self-aware and therefore not fully human; we are likely to act helplessly in accordance with uncontrolled inner drives or outer compulsions, like animals.

Mankind did not have to wait for the arrival of modern psychology to obtain teachings on these vitally important matters. Traditional wisdom, including all the great religions, has always described itself as "The Way" and given some kind of awakening as the goal. Buddhism has been called the "Doctrine of Awakening." Throughout the New Testament people are admonished to stay awake, to watch, not to fall asleep. At the beginning of the *Divine Comedy*, Dante *finds himself* in a dark wood, and he does not know how he got there, *"so full was I of slumber at that moment when I abandoned the true way."* It is not physical sleep that is the enemy of man; it is the drifting, wandering, shiftless moving of his attention that makes him incompetent, miserable, and less-than-fully-human. Without self-awareness, i.e., without a consciousness that is conscious of itself, man merely imagines that he is in control of himself, that he has free will and is able to carry out his intentions. In fact, as Ouspensky would put it, he has no more freedom to form intentions and act in accordance with them than has a machine. Only in occasional moments of self-awareness has he such freedom, and his most important task is by one means or another to make self-awareness continuous and controllable.

To achieve this, different religions have evolved different ways. The "heart of Buddhist meditation" is *satipatthana* or mindfulness. One of the outstanding Buddhist monks of today, Nyanaponika Thera, introduces his book on this subject with these words:

> This book is issued in the deep conviction that the systematic cultivation of Right Mindfulnes, as taught by the Buddha in his Discourse on Satipatthana, still provides the most simple and direct, the most thorough and effective, method for training the mind for its daily tasks and problems as well as for its highest aim: mind's unshakable deliverance from Greed, Hatred and Delusion. . . .
>
> This ancient Way of Mindfulness is as practicable today as it was 2,500 years ago. It is applicable in the lands of the West as in the

East; in the midst of life's turmoil as well as in the peace of the monk's cell.

The essence of the development of Right Mindfulness is an increase in the intensity and quality of attention, and the essence of *quality* of attention is its *bareness*.

> Bare attention is the clear and single-minded awareness of what actually happens *to* us and *in* us, at the successive moments of perception. It is called "bare", because it attends just to the bare facts of a perception as presented. . . . Attention or mindfulness is kept to a bare registering of the facts observed, without reacting to them by deed, speech or by mental comment which may be one of self-reference (like, dislike, etc.), judgement or reflection. If during the time, short or long, given to the practice of Bare Attention, any such comments arise in one's mind, they themselves are made objects of Bare Attention, and are neither repudiated nor pursued, but are dismissed, after a brief mental note has been made of them.[6]

These few indications may suffice to identify the essential nature of the method: Bare Attention is attainable only by stopping or, if it cannot be stopped, calmly observing all "inner chatter." It stands *above* thinking, reasoning, arguing, forming opinions—those essential yet subsidiary activities which classify, connect, and verbalize the insights obtained through Bare Attention. "In employing the methods of Bare Attention," says Nyanaponika, the mind "goes back to the seed state of things. . . . Observation reverts to the very first phase of the process of perception when mind is in a purely receptive state, and when attention is restricted to a bare noticing of the object."[7]

In the words of the Buddha: "In what is seen there must be only the seen; in what is heard there must be only the heard; in what is sensed (as smell, taste or touch) there must be only what is sensed; in what is thought there must be only what is thought."[8]

In short, the Buddha's Way of Mindfulness is designed to ensure that man's reason is supplied with genuine and unadulterated material before it starts reasoning. What is it that tends to adulterate the material? Obviously: man's egoism, his attachment to interests, desires, or, in Buddhist language, his Greed, Hatred, and Delusion.

Religion is the reconnection *(re-legio)* of man with reality, whether this Reality be called God, Truth, Allah, Sat-Chit-Ananda, or Nirvana.

The methods evolved in the Christian tradition are clothed, not surprisingly, in a very different vocabulary, but they nonetheless come to the same. Nothing can be achieved or attained as long as the little egocentric "I" stands in the way—there may, in fact, be many little, egocentric, and quite uncoordinated I's—and to get away from the "I," man must attend to "God," with "naked intent," as a famous English classic, *The Cloud of Unknowing,* calls it: "A naked intention directed to God, and himself alone, is wholly sufficient." The enemy is the intervention of *thought.*

> Should any thought arise and obtrude itself between you and the darkness, asking what you are seeking, and what you are wanting, answer that it is God you want: "Him I covet, him I seek, and nothing but him." . . . Quite possibly he [the thought] will bring to your mind many lovely and wonderful thoughts of his kindness. . . . He will go on chattering increasingly . . . [and] your mind will be well away, back in its old haunts. Before you know where you are you are disintegrated beyond belief! And the reason? Simply that you freely consented to listen to that thought, and responded to it, accepted it, and gave it its head.[9]

It is not a question of good or bad thoughts. Reality, Truth, God, Nirvana cannot be found by thought, *because thought belongs to the Level of Being established by consciousness* and not to that higher Level which is established by self-awareness. At the latter, thought has its legitimate place, but it is a subservient one. Thoughts cannot lead to *awakening* because the whole point is to awaken from thinking into "seeing." Thought can raise any number of questions; they may all be interesting, but their answers do nothing to wake us up. In Buddhism, they are called "vain thoughts": "This is called the blind alley of opinions, the gorge of opinions, the bramble of opinions, the thicket of opinions, the net of opinions." "Opinion, O disciples, is a disease; opinion is a tumor; opinion is a sore. He who has overcome all opinion, O disciples, is called a saint, one who knows."[10]

What is yoga? According to the greatest of yoga teachers, Patanjali (c. 300 B.C.), *Yoga is the control of the ideas in the mind.* Our circumstances are not merely the facts of life as we meet them, but also, and even more, the ideas in our minds. It is impossible to obtain any control over circumstances without first obtaining control over the ideas in one's mind, and the most important—as well as most universal—teaching of all the religions is that *vipassana* (to use a Buddhist term), clarity of vision, can be attained only by him who succeeds in putting the "thinking function" in its place, so that it maintains silence when ordered to do so and moves into action only when given a definite and specific task. Here is another quotation from *The Cloud of Unknowing:*

> Therefore the vigorous working of your imagination, which is always so active . . . must as often be suppressed. *Unless you suppress it, it will suppress you.*[11]

While the centerpiece of the Indian method is yoga, the centerpiece of the Christian method is prayer. To ask God for help, to thank Him, and to praise Him are legitimate purposes of Christian prayer, yet the essence of prayer goes beyond this. The Christian is called upon to "pray without ceasing." Jesus "spake a parable unto them to this end, that men ought always to pray, and not to faint."[12] This command has engaged the serious attention of Christians throughout the centuries. Perhaps the most famous passage on it is found in *The Candid Narrations of a Pilgrim to His Spiritual Father,* an anonymous jewel of world literature which was first printed in Russia in 1884.

> The first Epistle of St. Paul to the Thessalonians was read. In it we are exhorted, among other things, to *pray* incessantly, and these words engraved themselves upon my mind. I began to ponder whether it is possible to pray without ceasing, since every man must

occupy himself with other things needed for his support. . . . "What am I to do?" I mused. "Where will I be able to find someone who can explain it to me?"[13]

The pilgrim then obtains the *Philokalia*,[14] which "comprises the complete and minute knowledge of incessant inner prayer, as stated by twenty-five Holy Fathers."

This inner prayer is also called "the prayer of the heart"; while by no means unknown in the West, it has been brought to perfection mainly in the Greek and Russian Orthodox Churches. The essence of it is *"standing before God with the mind in the heart"*:

> The term "heart" is of particular significance in the Orthodox doctrine of man. When people in the west today speak of the heart, they usually mean the emotions and affections. But in the Bible, as in most ascetic texts of the Orthodox Church, the heart has a far wider connotation. It is the primary organ of man's being, whether physical or spiritual; it is the centre of life, the determining principle of all our activities and aspirations. As such, the heart obviously includes the affections and emotions, but it also includes much else besides: it embraces in effect everything that goes to comprise what we call a "person."[15]

Now, the *person* is distinguished from other beings by the mysterious power of self-awareness, and this power, as we have already noted, has its seat in the heart, where, in fact, it can be felt as a peculiar kind of warmth. The prayer of the heart, normally the Jesus Prayer (consisting, in English, of these twelve words: "Lord Jesus Christ, son of God, have mercy on me, a sinner") is endlessly repeated by the mind *in the heart*, and this vitalizes, molds, and reforms the whole person. One of the great teachers of this matter, Theophan the Recluse (1815–1894), explains thus:

> In order to keep the mind on one thing by the use of a short prayer, it is necessary to preserve attention and so lead it into the heart: for so long as the mind remains in the head, where thoughts jostle one another, it has no time to concentrate on one thing. But when attention descends into the heart, it attracts all the powers of

the soul and body into one point there. This concentration of all human life in one place is immediately reflected in the heart by a special sensation that is the beginning of future warmth. This sensation, faint at the beginning, becomes gradually stronger, firmer, deeper. At first only tepid, it grows into warm feeling and concentrates the attention upon itself. And so it comes about that, whereas in the initial stages the attention is kept in the heart by an effort of will, in due course this attention, by its own vigour, gives birth to warmth in the heart. This warmth then holds the attention without special effort. From this, the two go on supporting one another, and must remain inseparable; because dispersion of attention cools the warmth, and diminishing warmth weakens attention.[16]

The assertion that the endless repetition, silently, of a short sequence of words leads to a spiritual result, signalized, as it were, by *physical* sensations of *spiritual* warmth, is so strange to the modern mentality that it tends to be dismissed as mumbo-jumbo. Our pragmatism and respect for facts, of which we are so immensely proud, does not easily induce us to try it. Why not? Because trying it leads to the acquisition of certain insights, certain types of knowledge, which, once we have opened ourselves to them, will not leave us alone; they will present a kind of ultimatum: Either you change or you perish. The modern world likes matters it can trifle with, but the results of a direct approach to the study and development of self-awareness are not to be trifled with.

The First Field of Knowledge, in other words, is a minefield for anyone who fails to recognize that, at the human Level of Being, the *invisibilia* are of infinitely greater power and significance than the *visibilia*. To teach this basic truth has traditionally been the function of religion, and since religion has been abandoned by Western civilization, nothing remains to provide this teaching. Western civilization, consequently, has become incapable of dealing with the real problems of life at the human Level of Being. Its competence at the lower levels is breathtakingly powerful; but when it comes to the essentially human concerns, it is both ignorant and incompetent. Without the wisdom and disciplines of authentic religion the First Field of

Knowledge remains neglected, a wasteland overgrown with weeds, many of them poisonous. Healthy and useful plants may still appear there, but only accidentally. Without self-awareness (in the full sense of "factor z") man acts, speaks, studies, reacts mechanically, like a machine: on the basis of "programs" acquired accidentally, unintentionally, mechanically. He is not aware that he is acting in accordance with programs; it is therefore not difficult to reprogram him—to make him think and do quite different things from those he had thought and done before—provided only the new program does not wake him up. When he is awake, no one can program him: he programs himself.

This ancient teaching, which I am merely putting into modern terms, implies that there are two elements or agents involved rather than one: the computer programmer and the computer. The latter functions perfectly well without the attention of the former—as a machine. Consciousness—"factor y"—functions perfectly well without the presence of self-awareness, "factor z," as is demonstrated by all higher animals. That the fullness of the human "mind" cannot be accounted for by one element alone is the universal assertion of all the great religions, an assertion which has recently been corroborated by modern science. Just before his death at the age of eighty-four, Wilder Penfield, world-famous neurologist and brain surgeon, published a *summa* of his findings under the title *The Mystery of the Mind.* He says:

> Throughout my own scientific career I, like other scientists, have struggled to prove that the brain accounts for the mind. But now, perhaps, the time has come when we may profitably consider the evidence as it stands, and ask the question: *Do brain mechanisms account for the mind?* Can the mind be explained by what is now known about the brain? If not, which is the more reasonable of the two possible hypotheses. that man's being is based on one element, or on two?[17]

Dr. Penfield comes to the conclusion that "the mind seems to act independently of the brain in the same sense that a pro-

grammer acts independently of his computer, however much
he may depend upon the action of that computer for certain
purposes." He goes on to explain:

> Because it seems to me certain that it will always be quite impossible
> to explain the mind on the basis of neuronal action within the brain,
> and because it seems to me that the mind develops and matures
> independently throughout an individual's life as though it were a
> continuing element, and because a computer (which the brain is)
> must be operated by an agency capable of independent understand-
> ing, I am forced to choose the proposition that our being is to be
> explained on the basis of two fundamental elements.[18]

Obviously, the programmer is "higher" than the computer,
just as what I have called self-awareness is "higher" than consciousness. Studying the First Field of Knowledge implies the
systematic training of the "higher" factor. The programmer
cannot be trained simply by letting the computer run more
regularly or faster. His requirement is not simply knowledge of
facts and theories, but understanding or *insight*. Not surprisingly, the processes of gaining insight are quite different from
those of gaining factual knowledge. Many people are incapable
of seeing the difference between knowledge and insight and
therefore view methods of training like satipatthana, yoga, or
unceasing prayer as some kind of superstitious nonsense. Such
views are of course quite valueless and merely indicate a lack
of *adaequatio*. All systematic effort produces some kind of result.

> The Jesus Prayer acts as a constant reminder to make man look
> inwards *at all times,* to become aware of his fleeting thoughts, sud-
> den emotions and even movements so that it may make him try to
> control them. . . . By scrutinising and observing his own inner self
> he will obtain an increasing knowledge of his worthlessness which
> may fill him with despair. . . . These are the birth pangs of the spirit
> and the groanings of awakening spirituality in man. . . . One is
> advised to repeat the prayer of Jesus in "silence." . . . Silence here
> is meant to include inner silence; the silence of one's own mind, the
> arresting of the imagination from the ever-turbulent and ever-pre-

sent stream of thoughts, words, impressions, pictures and day-dreams, which keep one asleep. This is not easy, as the mind works almost autonomously.[19]

Few Western philosophers of the modern age have given serious attention to the methods of studying the First Field of Knowledge. A rare exception is W. T. Stace, for about twenty-five years, from 1935, a Professor of Philosophy at Princeton University. In his book on *Mysticism and Philosophy* he asks the long overdue question: "What bearing, if any, does what is called 'mystical experience' have upon the more important problems of philosophy?", and his investigations lead him to "the introvertive type of mystical experience," and thus to the methods employed by those seeking such experiences. It is perhaps unfortunate that Professor Stace uses the word "mystical," which has acquired a somewhat "mystical" meaning, when in fact nothing other is involved than the attentive exploration of one's own inner life. However, this does not detract from the pertinence and excellence of his observations.

First of all, he points out that *there is no doubt* that the basic psychological facts about this "introvertive experience" are in essence "the same all over the world in all cultures, religions, places, and ages." Professor Stace writes as a philosopher and does not claim to have any personal experience of these matters. He therefore finds them very strange indeed. "They are," he says, "so extraordinary and paradoxical that they are bound to strain belief when suddenly sprung upon anyone who is not prepared for them." He then proceeds to set forth "the alleged facts as the mystics state them without comment and without passing judgment." Although he states the facts in terms which no mystic has ever used, his method of exposition is so clear that it is worth reproducing:

> Suppose that one should stop up the inlets of the physical senses so that no sensations could reach consciousness. . . . There seems to be no *a priori* reason why a man bent on the goal . . . should not, by acquiring sufficient concentration and mental control, exclude all physical sensations from his consciousness.

Suppose that, after having got rid of all sensations, one should go on to exclude from consciousness all sensuous images and then all abstract thoughts, reasoning processes, volitions, and other particular contents; what would then be left of consciousness? There would be no mental content whatever but rather a complete emptiness, vacuum, void.[20]

This is, of course, precisely the aim pursued by those who wish to study their inner life: the exclusion of all disturbing influences emanating from the senses or from the "thinking function." Professor Stace, however, becomes deeply puzzled:

One would suppose *a priori* that consciousness would then entirely lapse and one would fall asleep or become unconscious. But the introvertive mystics—thousands of them all over the world—unanimously assert that they have attained to this complete vacuum of particular mental contents, but what then happens is quite different from a lapse into unconsciousness. On the contrary, what emerges is a state of *pure* consciousness—"pure" in the sense that it is not the consciousness *of* any empirical content. It has no content except itself.[21]

In the language I used previously we might say: the computer programmer emerges, who, of course, has none of the "contents" of the computer; in other words again: the factor z—self-awareness—really comes into its own when, and only when, the factor y—consciousness—leaves the center of the stage.

Professor Stace says: "The paradox is that there should be a positive experience which has no positive content—an experience which is both something and nothing." But there is nothing paradoxical in a "higher" force displacing a "lower" force, in an experience which is something but *no thing*. The paradox exists only for those who insist on believing that there can be nothing "higher than" or "above" their everyday consciousness and experience. How can they believe such a thing? Everybody, surely, has had some moments in his life which held more significance and realness of experience than his everyday life. Such moments are pointers, glimpses of unrealized potentialities, flashes of self-awareness.

Professor Stace continues his exploration thus:

Our normal everyday consciousness always has objects, or images, or even our own feelings or thoughts perceived introspectively. Suppose then that we obliterate all objects physical or mental. *When the self is not engaged in apprehending objects it becomes aware of itself.* The self itself emerges. . . . One may also say that the mystic gets rid of the empirical ego whereupon the pure ego, normally hidden, emerges into the light. *The empirical ego is the stream of consciousness. The pure ego is the unity which holds the manifold of the stream together.*[22]

The essential identity of these views with those of Wilder Penfield is unmistakable. Both corroborate the central teaching of the great religions, which, in many different languages and modes of expression, urge man to open himself to the "pure ego" or "Self" or "Emptiness" or "Divine Power" that dwells within him; to awaken, as it were, out of the computer into the programmer; to transcend consciousness by self-awareness. Only by liberating oneself from the thralldom of the senses and the thinking function—both of them servants and not masters—by withdrawing attention from *things seen* to give it to *things unseen* can this "awakening" be accomplished. "We look not at the things which are seen, but at the things which are not seen; for the things which are seen are temporal; but the things which are not seen are eternal."[23]

There is a very great deal more that could be said about this greatest of all arts, the acquisition of self-knowledge, by, in our terminology, the study of the First Field of Knowledge. It will be more useful, however, to turn now to the *Second Field of Knowledge,* that is, the knowledge we may obtain of the *inner* experience of *other* beings. One thing is certain: We seem to have no *direct access* to such knowledge. How, then, is such knowledge possible at all?

7

The Four Fields of Knowledge: 2

The higher the Level of Being, the greater is the importance of inner experience, i.e., the "inner life," as compared with outer appearance, i.e., such directly observable and measurable attributes as size, weight, color, movement, etc.; also, the more likely are we to be able to obtain some knowledge of the "inner life" of other beings, at least up to the human level. We are convinced that we can indeed know something of what goes on inside another human being, a little bit even about the inner life of animals, virtually nothing about that of plants, and certainly nothing at all about that of stones and other pieces of inanimate matter. When Saint Paul says that "we know that the whole creation groaneth and travaileth in pain together until now,"[1] we can glimpse his meaning with regard to people and possibly animals but have very great difficulties when it comes to plants and minerals.

Let us then begin with other people: How do we gain knowledge of what is going on inside them? As I have said before, we are living in a world of *invisible* people; most of them do not even wish us to know anything about their inner life; they say, "Don't intrude, leave me alone, mind your own business." Even when somebody wants to "bare his soul" to someone else, he finds it extraordinarily difficult to do so, he does not know how to express himself, and, without in the least wishing to mislead, he tends to say many things that are not true at all. In desperation he may try to communicate without words, by gestures, signs, bodily touch, shouting, weeping, even violence.

Although there are constant temptations to forget it, we all know that our lives are made or marred by our relationships with other human beings; no amount of wealth, health, fame, or power can compensate us for our loss if these relationships go wrong. Yet they all depend on our ability to *understand* others and their ability to *understand* us.

Most people seem to believe that there is nothing more to this problem of communication than listening to another person's speech and observing the outward movements of his body, in other words, that we can rely implicitly on other people's *visible* signals to convey to us a correct picture of their *invisible* thoughts, feelings, intentions, and so on. Alas! the matter is not as simple as this. Consider the requirements step by step, assuming that there is a genuine wish on the part of one person to convey his thought to another person (and leaving out all possibilities of deliberate deception).

First, the communicant must know, with some precision, what the thought *is* he wishes to convey.

Second, he must find visible (and audible) symbols—gestures, bodily movements, words, intonation, etc.—which in his judgment most accurately "externalize" his "internal" thought. This may be called *"the first translation."*

Third, the listener must have a faultless reception of these visible (etc.) symbols, which means that he must not only accurately hear what is being said but also accurately observe the nonverbal symbols (such as gesture and intonation) that are being employed.

Fourth, the listener must then in some way integrate the numerous symbols he has received and turn them back into thought. This may be called *"the second translation."*

It is not difficult to see that much can go wrong at each stage of this four-stage process, particularly with the two "translations." In fact, we might come to the conclusion that reliable and accurate communication is impossible. Even if the communicant is completely clear about the thought he wishes to convey, his choice of symbols—gestures, word combinations, intonation—is a highly personal matter; and even if the recipient listens and observes perfectly, how can he be sure that he attaches the appropriate meaning to the symbols he receives? These doubts and questions are only too well justified. The process described is extremely laborious, and unreliable even when immense time and effort are spent to formulate definitions, explanations and exceptions, provisos and escape clauses. We are immediately reminded of legal or international diplomatic documents. This, we might think, is a case of communication between two "computers," where everything has to be reduced to pure logic: either/or. Here the dream of Descartes becomes real: nothing counts except precise, distinct, and absolutely certain ideas.

Yet, miraculously, in real life perfect communication *is* possible and not infrequent. It proceeds without elaborate definitions or provisos or escape clauses. People are even capable of saying: "I don't like the way you are putting it, but I agree with what you mean." This is highly significant. There can be a "meeting of minds," for which the words and gestures are little more than an invitation. Words, gestures, intonation—these can be one of two things (or even a bit of each): computer language or an invitation to two "computer programmers" to get together.

If we cannot achieve a real "meeting of minds" with the people nearest to us in our daily lives, our existence becomes an agony and a disaster. In order to achieve it, I must be able to gain knowledge of what it is like to be "you," and "you" must be able to gain knowledge of what it is like to be me. Both of us must become knowledgeable in what I call the Second Field of Knowledge. Since we know that very little knowledge comes naturally to most of us and that the acquisition of greater knowledge requires effort, we are bound to ask ourselves the question: "What can I do to acquire greater knowledge, to become more understanding of what is going on inside the people with whom I live?"

Now, the remarkable fact is that all traditional teachings give one and the same answer to this question: "You can understand other beings only to the extent that you know yourself." Naturally, close observation and careful listening are necessary; the point is that even perfect observation and perfect listening lead to nothing unless the *data* thus obtained are correctly interpreted and understood, and the precondition to my ability to understand correctly is my own self-knowledge, my own inner experience. In other words, and using our previous terminology: there must be *adaequatio*, item by item, bit by bit. A person who had never consciously experienced bodily pain could not possibly know anything about the pain suffered by others. The outward signs of pain—sounds, movements, a flow of tears—would of course be noticed by him, but he would be totally *inadequate* to the task of understanding them correctly. No doubt he would attempt some kind of interpretation; he might find them funny or menacing or simply incomprehensible. The *invisibilia* of the other being—in this case the inner experience of pain—would remain invisible to him.

I leave it to the reader to explore the enormous range of inner experiences which fill the lives of men and women. As I have emphasized before, they are all invisible, inaccessible to external observation. The example of bodily pain is instructive precisely because there is no subtlety about it. Few people doubt the reality of pain, and the realization that here is a thing we all recognize as real, true, one of the great "stubborn facts" of human existence, which nonetheless is *unobservable by our outer senses*, may come as a shock. If only that which can be observed by our outer senses is deemed to be real, "objective," scientifically respectable, pain must be dismissed as unreal, "subjective," unscientific. And the same applies to everything else that moves us internally: love and hatred, joy and sorrow, hope, fear, anguish, and so on. If all these forces or movements inside me are not *real*, they need not be taken seriously, and if I do not take them seriously in myself, how can I consider them real and take them seriously in another being? It is, in fact, more convenient to assume that other beings do not really suffer as we do and do not really possess an inner life as complex, subtle, and vulnerable as our own. Indeed, throughout the ages man has shown an enormous capacity to bear the sufferings of others with fortitude and equanimity. Since, moreover, as J. G. Bennett has shrewdly observed,[2] we tend to see ourselves primarily in the light of our intentions, which are invisible to others, while we see others mainly in the light of their actions, which are visible to us, we have a situation in which misunderstanding and injustice are the order of the day.

There is no escape from this situation except by the diligent and systematic cultivation of the First Field of Knowledge, through which—and through which alone—we can obtain the insights needed for the cultivation of the Second Field of Knowledge, i.e., knowledge of the inner experiences of beings other than ourselves.

To be able to take the inner life of my neighbor seriously, it is necessary that I take my own inner life seriously. But what does that mean? It means that I must put myself in a condition where I can truly observe what is going on and begin to understand what I observe. In modern times there is no lack of understanding of the fact that man is a *social* being and that "No man is an Iland, intire of it selfe" (John Donne, 1571–1631). Hence there is no lack of exhortation that he should love his neighbor —or at least not be nasty to him—and should treat him with tolerance, compassion, and understanding. At the same time, however, the cultivation of self-knowledge has fallen into virtually total neglect, except, that is, where it is the object of active suppression. That you cannot love your neighbor unless you love yourself; that you cannot understand your neighbor unless you understand yourself; that there can be no knowledge of the "invisible person" who is your neighbor except on the basis of self-knowledge—these fundamental truths have been forgotten even by many of the professionals in the established religions.

Exhortations, consequently, cannot possibly have any effect; genuine understanding of one's neighbor is replaced by sentimentality, which, of course, crumbles into nothingness as soon as self-interest is threatened and fear of any kind is aroused. Knowledge is replaced by assumptions, trite theories, fantasies. The enormous popularity of the crudest and meanest psychological and economic doctrines, purporting to "explain" the actions and motives of others—never of ourselves!—shows the disastrous consequences of the current lack of competence in the Second Field of Knowledge, which, in turn, is the direct result of the modern refusal to attend to the First Field of Knowledge, self-knowledge.

Anyone who goes openly on a *journey into the interior,* who withdraws from the ceaseless agitation of everyday life and pursues the kind of training—satipatthana, yoga, Jesus Prayer, or something similar—without which genuine self-knowledge cannot be obtained, is accused of selfishness and of turning his back on his social duties. Meanwhile, world crises multiply and everybody deplores the shortage, or even total lack, of "wise" men or women, unselfish leaders, trustworthy counselors, etc. It is hardly rational to expect such high qualities from people who have never done any *inner work* and would not even understand what was meant by the words. They may consider themselves decent, law-abiding people and good citizens; perhaps "humanists," even "believers." It makes very little difference how they *dream about themselves.* Like a "pianola" they play mechanical music; like a computer they carry out prearranged programs. The programmer is asleep.

An important part of the modern "program" is to reject religion as cheaply moralizing, outdated, ceremonial dogmatism, thereby rejecting the very force, perhaps the only force, that could wake us up and lift us to the truly human level, that of self-awareness, self-control, self-knowledge, and, thereby, knowledge and understanding of others; and which would give us the *power to help them* when necessary.

People say: It is all a matter of communication. Of course it is. But communication, as we have said, implies two "translations"—from thought to symbol and from symbol to thought. Symbols cannot be understood like mathematical formulae;

they have to be experienced *interiorly*. They cannot properly be taken up by consciousness, but only by self-awareness. A gesture, for instance, cannot be understood by the rational mind; we have to become aware of its meaning inside ourselves, with our body rather than with our brain. Sometimes the only way to understand the mood or feelings of another person is by imitating his posture, gestures, and facial expressions. There is a strange and mysterious connection between the interior-invisible and the exterior-visible. William James was interested in the bodily expression of emotions and advanced the theory that the emotion we feel is nothing but the feeling of some bodily changes:

> Common-sense says, we lose our fortune, are sorry and weep; we meet a bear, are frightened and run; we are insulted by a rival, are angry and strike. The hypothesis here to be defended says that this order of sequence is incorrect . . . and that the more rational state-ment is that we feel sorry because we cry, angry because we strike, afraid because we tremble, and not that we cry, strike, or tremble, because we are sorry, angry, or fearful, as the case may be.[3]

The hypothesis, although probably more remarkable for its originality than for its truth value, brings into sharp focus the intimate connection between inner feeling and bodily expres-sion; it points to a mysterious bridge connecting the invisible and the visible, and identifies the body as an *instrument of knowledge*. I have no doubt that a baby learns a great deal about its mother's emotions by imitating her posture and facial movements and then discovering what feelings are associated with these bodily expressions.

It is for these reasons that all methods devised for the acquisi-tion of self-knowledge (Field 1) pay a great deal of attention to bodily postures and gestures, for the establishment of control over the body is, to say the least, the *first step* in the establish-ment of control over the thinking function. Uncontrolled agita-tion of the body inevitably produces uncontrollable agitation of the mind, a condition which precludes all serious study of one's inner world.

When a high degree of inner calm and quietude has been established, the "computer" is left behind and the "computer programmer" comes into his own. In Buddhist terms, this is called *vipassana,* "clarity of vision." In Christian terms, we say there is some kind of encounter with a higher Level of Being, above the human level. Naturally, those of us who have no personal experience of this higher level cannot imagine it, and the language of those who are trying to tell us about it either means nothing to us or suggests a disordered mind, even madness. We have no easy means of distinguishing between infrahuman and suprahuman "madness." But we can look at the whole life of the person in question: If it displays evidence of great intellectual powers, organizing ability, wisdom, and personal influence, we can be quite certain, when we cannot understand it, that

> The fault, dear Brutus, is not in our stars,
> But in ourselves, that we are underlings.

No one is *adequate* to that which lies above him. However, an inkling and intimation of possibilities and an inspiration toward a real effort of awakening can always be obtained.

There is today a great deal of talk about the attainment of "higher states of consciousness." Unfortunately, this aspiration, in most cases, does not grow out of a deep respect for the great wisdom traditions of mankind, the world religions, but is based on such fantastic notions as an "Aquarian Frontier" or the "Evolution of Consciousness," and is generally associated with a total inability to distinguish between the spiritual and the occult. It seems that the real aim of these movements is to obtain new thrills, to master magic and miracles, thereby enlivening existential boredom. The advice of all people knowledgeable in these matter is *not* to seek occult experiences and *not* to pay any attention to them when they occur—and they will almost inevitably occur when any intensive inner work is undertaken. The great teacher of Buddhist Satipatthana Meditation, the Venerable Mahasi Sayadaw (1904–1955), warns the pupil that he will have all sorts of extraordinary experiences:

A brilliant light will appear to him. To one it will appear like the light of a lamp, to others like a flash of lightning, or like the radiance of the moon or the sun, and so on. With one it may last for just one moment, with others it may last longer. . . . There arises also *rapture . . . tranquillity* of mind . . . a very sublime feeling of *happiness.* . . . Having felt such rapture and happiness accompanied by the "brilliant light" . . . the meditator now believes: "Surely I must have attained to the Supra-mundane Path and Fruition! Now I have finished the task of meditation." This is mistaking what is not the Path for the Path, and it is a corruption of Insight which usually takes place in the manner just described. . . . After noticing these manifestations of Brilliant Light and the others, or after leaving them unheeded, he [the true seeker] goes on continuously as before . . . he gets over the corruptions relating to brilliant light, rapture, tranquillity, happiness, attachment, etc.[4]

Christian saints and sages are equally clear on this point. We can take Saint John of the Cross (1542–1591) as a typical example:

With respect to all [bodily senses] there may come, and there are wont to come, to spiritual persons representations and objects of a supernatural kind. . . .

And it must be known that, although all these things may happen to the bodily senses in the way of God, *we must never rely upon them or accept them,* but must always fly from them, without trying to ascertain whether they be good or evil . . . for the bodily sense is as ignorant of spiritual things as is a beast of rational things, and even more so.

So he that esteems such things *errs greatly and exposes himself to great peril of being deceived;* in any case he will have within himself a *complete impediment to the attainment of spirituality.*[5]

The New Consciousness which is so much talked about today cannot help us out of our difficulties and will merely increase the prevailing confusion unless it arises from a genuine search for self-knowledge (the First Field of Knowledge) and moves on to an equally genuine study of the inner life of other beings (the Second Field of Knowledge) and also to the Third Field of Knowledge (which will be discussed later). If it leads merely to fascination with occult phenomena, it belongs to the Fourth

Field of Knowledge (also to be considered later), and can do nothing to improve our understanding of ourselves and of our fellow creatures.

Inner work, or *yoga* in its many forms, is not a peculiarity of the East, but the taproot, as it were, of all authentic religions. It has been called "the applied psychology of religion,"[6] and it must be said that *religion without applied psychology is completely worthless.* "Simply to believe a religion to be true, and to give intellectual assent to its creed and dogmatic theology, and not to know it to be true through having tested it by the scientific methods of *yoga,* results in the blind leading the blind."[7] This statement comes from W. Y. Evans-Wentz, who spent most of his life "editing" sacred writings from Tibet and making them available to the West. He asks:

> Is Occidental man for much longer to be content with the study of the external universe, and not know himself? If, as the editor believes, the Oriental sage is able to direct us of the Occident to a method of attaining scientific understanding of the hidden side of man's nature, are we not unwise in failing to give it unprejudiced scientific examination?
>
> Applied sciences in our portion of the world are, unfortunately, limited to chemistry, economics, mathematics, mechanics, physics, physiology, and the like; and anthropology and psychology as applied sciences in the sense understood in *yoga* are for almost all Occidental scientists mere dreams of impracticable visionaries. We do not believe however, that this unsound view can long endure.[8]

"Applied sciences in the sense understood in *yoga*" means a science that finds its material for study not in the appearances of other beings but *in the inner world of the scientist himself.* This inner world, of course, is not worth studying—and nothing can be learned from it—if it is an impenetrable chaos. While the methods of Western science can be applied by anyone who has learned them, the scientific methods of yoga can be effectively applied only by those prepared first of all to put their own house in order through discipline and systematic inner work.

Self-knowledge is not only the precondition of understanding other people; it is also the precondition of understanding, at least to some extent, the inner life of beings at lower levels: animals and even plants. Saint Francis could communicate with animals, and so could other men and women who had attained an exceptional degree of self-mastery and self-knowledge. Reverting to our earlier way of speaking, we can say: Such communicating is not possible for the computer, but only for the computer programmer. His powers certainly go far beyond those we are ordinarily familiar with and are not confined to the framework of time and space.

Ernest E. Wood, who really could speak from experience of yoga, says: "I wish to guard the novice against the two dangers of self-judgment and the fixing of goals, and to tell him that as he is calling into expression high forces within and behind and above his present level of self he must let them do their work in him."[9] It is therefore neither necessary nor advisable to talk about these matters in detail. Those who are genuinely interested—not in the attainment of powers but in their own inner development—will study the lives and works of people who have put themselves under the control of "Higher Mind" and thus broken out of our ordinary confinement of time and space. There is no lack of examples from all ages and all parts of the world. It will serve our present purposes to have a quick look at three recent cases where the higher possibilities of the human being have manifested themselves almost under our very eyes.

The first case is that of Jakob Lorber, who was born in Styria, a province of Austria, in 1800. His father owned two small vineyards which produced a meager living for the family but was also a musician who could play virtually all instruments and was able to earn some extra income as a conductor. His eldest son, Jakob, learned to play the organ, piano, and violin and showed exceptional musical talent but had to wait until his fortieth year before he was offered an appointment that promised to give him scope for his talents. He was on the point of leaving Graz to take up his new job at Trieste when he heard inside himself a very clear voice ordering him to "get up, take a pen, and write." This was on March 15, 1840, and Jakob Lorber stayed at Graz and wrote down what the inner voice dictated to him until he died, aged sixty-four, on August 24, 1864. During these twenty-four years, he produced the equivalent of twenty-five volumes of four hundred pages each, a monumental "New Revelation." The original manuscripts are still in existence, and they show an absolutely even flow of writing, with hardly any corrections. Many prominent men of his time were intimate friends of Lorber's; some of them supported him with food and money during the years of his writing activity, which left him little time for earning a living. A few have written down their impressions of this humble and totally unpretentious man, who lived in poverty and often experienced his writing task as a very heavy burden.

The centerpiece of Lorber's writings is the *New St John's Gospel* in ten large volumes. I shall not attempt here to describe or in any way to characterize these works, all written in the first person singular: "I, Jesus Christ, am speaking." They contain many strange things which are unacceptable to the modern mentality, but at the same time such a plethora of high wisdom and insight that it would be difficult to find anything more impressive in the whole of world literature. At the same time, Lorber's books are full of statements on scientific matters which flatly contradict the sciences of his time and anticipate a great deal of modern physics and astronomy. No one has ever raised the slightest doubt that the Lorber manuscripts came into existence during the years 1840–1864 and were produced by Jakob Lorber alone. There is no rational explanation for the range, profundity, and precision of their contents. Lorber himself always assured, and evidently convinced, his friends that none of it flowed from his own mind and that no one was more astonished at these contents than he himself.[10]

The case of Edgar Cayce is perhaps even more striking. He lived in the United States from 1877 to 1945 and left well over fourteen thousand stenographic records of statements he made during a kind of sleep, answering very specific questions from over six thousand different people, in the course of forty-three years. These statements, generally referred to as "readings," constitute "one of the largest and most impressive records of psychic perception ever to emanate from a single individual. Together with their relevant records, correspondence and reports, they have been cross indexed under thousands of subject headings and placed at the disposal of psychologists, students, writers and investigators who still come, in increasing numbers, to examine them."[11]

Like Jakob Lorber, Edgar Cayce lived modestly, even in poverty, for many years of his life. He certainly never exploited the immense fame he gained during his lifetime. The work his gifts imposed upon him was all too often a heavy burden on him, and, although short-tempered, he never lost his modesty and simplicity. Thousands of people asked him for medical help. Putting himself into some kind of trance, he was able to give generally accurate diagnoses of the illnesses of complete strangers living hundreds or even thousands of miles away. "Apparently," he said, "I am one of the few who can lay aside their own personalities sufficiently to allow their souls to make this attunement to the universal source of knowledge—but I say this without any desire to brag about it. . . . I am certain all human beings have much greater powers than they are ever conscious of—if they would only be willing to pay the price of detachment from self-interest that it takes to develop those abilities. *Would you be willing, even once a year, to put aside, pass out entirely from, your own personality?*"[12]

Even more contemporary than Edgar Cayce is Therese Neumann, also known as Therese of Konnersreuth, who lived in southern Germany from 1898 to 1962. Much can be related of Therese's inner life and its extraordinary outward manifestations, but perhaps the most noteworthy feature is this: This sturdy, cheerful, immensely common-sensical peasant woman lived for thirty-five years without ingesting any liquid or food except the daily Eucharist. This is not a legend from a remote place or time; it happened under our eyes, observed by innumerable people, investigated virtually continuously for thirty-five years, at Konnersreuth in what was called the American Zone of Western Germany.

Jakob Lorber, Edgar Cayce, and Therese Neumann were intensely religious personalities who never ceased to aver that all their knowledge and power came from "Jesus Christ"—a level infinitely above their own. At this suprahuman level, each of them found, in their various ways, liberation from constraints that operate at the level of ordinary humanity—limits imposed by space and time, by the needs of the body, and by the opaqueness of the computer-like mind. All three examples illustrate the paradoxical truth that such "higher powers" cannot be acquired by any kind of attack and conquest conducted by the human personality; only when the striving for "power" has entirely ceased and been replaced by a certain transcendental longing, often called the love of God, may they, or may they not, be "added unto you."

8

The Four Fields of Knowledge: 3

In the face of facts such as those presented by the lives of Jakob Lorber, Edgar Cayce, Therese Neumann, and indeed countless others, the modern world abandons its pragmatic attitudes, of which it is so proud, and simply shuts its eyes, for it has a methodical aversion to anything pertaining to a Level of Being that is higher than that of the most humdrum and ordinary life.

This aversion is not untinged by fear. Are there not great dangers in the pursuit of self-knowledge? There are indeed—and this takes us to a consideration of the *Third Field of Knowledge.* The systematic study of the inner worlds of myself (Field 1) and of other beings (Field 2) *must* be balanced and complemented by an equally systematic study of *myself as an objective phenomenon.* Self-knowledge, to be healthy and complete, *must* consist of two parts—knowing my own inner world (Field 1) and "knowing myself as I am known" by others (Field 3). Without the latter, the former may indeed lead to the grossest and most destructive delusions.

We have direct access to Field 1, but no direct access to Field 3. As a result, our intentions tend to be much more real to us than our actions, and this can lead to a great deal of misunderstanding with other people, to whom our actions tend to be much more real than our intentions. If I derive my "picture of myself" solely from Field 1, my inner experiences, I inevitably tend to see myself as "the center of the Universe": Everything revolves around me; when I shut my eyes, the world disappears; my suffering turns the world into a vale of tears; my happiness turns it into a garden of delight. A passage from the diaries of Goebbels—one of the Big Three of Hitler's Germany—comes to mind: "If we perish," he says, "the whole world will perish." But we do not need such gruesome examples. There are harmless and mild-mannered philosophers who raise the question whether the tree at which they are gazing will still be there when nobody is looking. They have lost themselves in Field 1 and have not been able to reach Field 3.

In Field 3, totally detached, objective observation is required, unalloyed by any wishful associations. What do I really observe? Or rather: What would I see if I could see myself as I am seen? Achieving this is a very difficult task. Unless it is accomplished, harmonious relationships with other people are impossible, and unless I am aware of my actual impact upon others, the injunction "Don't do to others what you don't want them to do to you" becomes meaningless.

I once read a story of a man who died and went into the next world where he met numbers of people some of whom he knew and liked and some he knew and disliked. But there was one person there whom he did not know and he could not bear him. Everything he said infuriated and disgusted him—his manner, his habits, his laziness, his insincere way of speaking, his facial expressions—and it seemed to him also that he could see into this man's thoughts and his feelings and all his secrets and, in fact, into all his life. He asked the others who this impossible man was. They answered: "Up here we have very special mirrors which are quite different from those in your world. This man is yourself." Let us suppose, then, that you have to live with a person who is you. Perhaps this is what the other person has to do. Of course, if you have no self-observation you may actually imagine this would be charming and that if everyone were just like you, the world would indeed be a happy place. There are no limits to vanity and self-conceit. Now in putting yourself into another person's position you are also putting yourself into his point of view, into *how* he sees you, and hears you, and experiences you in your daily behaviour. You are seeing yourself through his eyes.[1]

This is a vivid and accurate description of what it means to obtain knowledge in Field 3, and it incidentally makes quite clear that knowledge in Field 1 is very different from knowledge in Field 3, and that the former without the latter may be worse than useless.

Everybody has a very natural curiosity as to what he looks like, what he sounds like, and what impression he makes on others. But the "very special mirrors" of the story do not exist on this earth, perhaps mercifully so. The shocks they would administer might be more than we could take. It is always painful to realize that there really *is* quite a lot wrong with oneself, and we possess many mechanisms to protect ourselves from this revelation. Our natural curiosity, therefore, does not take us very far into Field 3, and we are all too easily diverted into studying the faults of others rather than our own. Maurice Nicoll reminds us of the words in the Gospels: "Why beholdest thou the mote that is in thy brother's eye, but considerest not the beam that is in thine own eye?" "In the Greek," he points out, "the word used for the mote is simply *see*. That is easy to do. But the word used for the beam in oneself is interesting. It means 'to take notice of, to detect, to acquire knowledge of, to take in a fact about, to learn, to observe, to understand'. Obviously something far more difficult is meant than merely seeing another's faults. To turn round is not easy."[2]

How, then, can we fulfill this task, so crucial for the harmony of our life with others? The methodology is set out in the books of traditional religions, albeit in a scattered form. Perhaps the most helpful guidance in this field is to be found in Nicoll's *Psychological Commentaries on the Teaching of Gurdjieff and Ouspensky,* from which I have just quoted. His guidance goes under the term "External Considering," or putting yourself into the other person's place. This requires a very high degree of inner truthfulness and freedom. It cannot be learned in a day, and good intentions cannot succeed without protracted effort.

What kind of effort? Nothing in this line is possible without *self-awareness.* To put myself into another person's situation, I must detach myself from my own situation. Mere consciousness will not do so; it only confirms me in my own situation. The computer can do nothing but carry out its pre-established program. Only the computer programmer can effect a real change, such as "putting oneself into another person's situation." In other words, the quality or power required is not simply *consciousness*—what I have called "factor y," which enables beings to be animals—but *self-awareness,* "factor z," which enables animals to be human beings. As Nicoll puts it, "External considering is very good work. It is not about whether you were right or the other person. It increases consciousness,"[3] and I would add: *"to the level of self-awareness."*

One of the things we are least aware of in ourselves is our own "swing of the pendulum." Other people notice how we contradict ourselves, but we do not. Knowledge in Field 3, by enabling us to see ourselves as others see us, will help us to see our contradictions. This is a matter of quite fundamental importance, as we shall see later. It is not as if apparent contradictions were necessarily manifestations of error; more likely, they are manifestations of Truth. Opposites coexist throughout reality, but we always find it difficult to keep two opposites in our mind at the same time. Others can easily observe the swing of my pendulum from one opposite to the other, just as I can easily observe the swing of theirs. But it is my task—in Field 3—to become fully aware of the swing of *my* pendulum, of the fact that *I* tend to change very often from one opinion to its opposite; and it is my task not merely to notice the change but to take note of it uncritically, without judging or justifying it. The essence of the task in Field 3 is *uncritical self-observation,* so that we obtain cool, objective pictures of what is actually happening, not pictures "retouched" by our current opinions of right or wrong.

One of the methods of study in Field 3 is "taking photographs," that is, catching true glimpses of oneself, as sometimes happens when we are not aware of looking at ourselves. Nicoll has this to say:

> If you have taken an album of good photographs of yourself through long self-observation, then you will not have to look far in it to find in yourself what you object to so much in the other person, and then you will be able to put yourself in the other person's position, to realise that he has also this thing that you have noticed in yourself, that he has his inner difficulties, just as you have, and so on.
>
> The less vanity . . . you have, and the more you externally consider, the less important will you think yourself.[4]

While the—necessary!—studies in Field 1 may tend to raise one's feelings of self-importance, the counterbalancing studies in Field 3 should lead to the realization of one's nothingness. What am I in this great, great Universe? What am I? Just one little ant among four thousand millions of them on the face of this puny little Earth! In the words of Pascal, "Man is only a reed, the weakest thing in nature; but he is a thinking reed"— that is, a reed with self-awareness, and to that extent infinitely precious, even if, most of the time, his self-awareness remains a mere dormant potentiality.

Our main help in obtaining knowledge in Field 3 comes from the fact that we are *social* beings; we live not alone but with others. And these others are a kind of mirror in which we can see ourselves as we actually are, not as we imagine ourselves to be. The best way to obtain the requisite knowledge about ourselves, therefore, is to observe and understand the needs, perplexities, and difficulties of others, putting ourselves in their situation. One day we may get to the point when we can do this so perfectly that *we*, little "egos" with their own needs, perplexities, and difficulties, do not come into this picture at all. Such total absence of *ego* would mean total objectivity and total effectiveness.

The Christian is told "to love his neighbor as himself." What does that mean? When a person loves himself there is nothing standing between the giver and the receiver of that love. But when he loves his neighbor his own little *ego* tends to stand in between. To love one's neighbor as one loves oneself, therefore, means to love without any interference from one's own ego; it means the attainment of perfect altruism, the elimination of all traces of egoism.

Just as compassion is the prerequisite of learning in the Second Field of Knowledge, so altruism is the prerequisite of learning in the Third.

We have noted before that these two fields are not "directly accessible" to our observation. Only through the higher qualities of compassion and altruism are we able to enter them.

9

The Four Fields of Knowledge: 4

We turn now to a consideration of the *Fourth Field of Knowledge,* the "appearance" of the world around us. By "appearance" I mean everything that offers itself to our senses. In the Fourth Field of Knowledge the decisive question is always "What do I actually observe?" and progress is attained by eliminating assumptions, notions, presuppositions as to causes, etc., which cannot be verified by sense observation. Field 4, therefore, is the real homeland of every kind of *behaviorism:* only strictly observable *behavior* is of interest. All the sciences are busy in this field, and many people believe that it is the only field in which true knowledge can be obtained.

As an example, we may quote Vilfredo Pareto (1848–1923), whose *Trattato di Sociologia Generale* has been hailed as "the greatest and noblest effort" ever undertaken in the direction of "objective thinking without sentiment, and . . . the methods by which the rational state of mind can be cultivated."[1] Pareto, like countless others, insists that only in what I call "Field 4" can there be a "scientific approach":

> The field in which we move is therefore the field of experience and observation strictly. We use those terms in the meanings they have in the natural sciences such as astronomy, chemistry, physiol-

ogy, and so on, and not to mean those other things which it is the fashion to designate by the terms "inner" or "Christian" experience.[2]

Pareto, in other words, wishes to base himself exclusively on "experience and observation," and he restricts the meaning of these terms to facts which the outer senses, aided by instruments and other apparatus and guided by theories, can ascertain. He thereby excludes all *inner* experiences, like love and hate, hope and fear, joy and anguish, and even pain. This he considers the only rational approach, and a recipe for real success:

> One readily understands how the history of the sciences down to our own time is substantially a history of the battles against the methods of introspection, etymology, analysis of verbal expression. . . . In our day the [latter] method has been largely banished from the physical sciences, *and the advances they have made are the fruit of that proscription.* But it is still strutting about in political economy and more blatantly still in sociology; whereas if those sciences would progress, it is imperative that they should follow *the example set by the physical sciences.*[3]

Here, it is clear that Pareto is unwilling or unable to distinguish between the different Levels of Being. It is one thing to banish "inner" knowledge from the study of inanimate nature, the lowest of the four Levels of Being, simply because, as far as we know, there is no inner life at this level, and everything is "appearance." It is quite another thing to banish it from the study of human nature and behavior, at the highest of the four Levels of Being, where outer appearance is a very unimportant matter compared with inner experience.

In the Second Field of Knowledge—the inner experience of other beings—we found that we can know most about the higher levels and least about inanimate matter. In the Fourth Field of Knowledge, it is the other way round: We can know most about inanimate matter and least about human beings.

From Pareto's point of view, "There is not the slightest difference between the laws of political economy or sociology and the laws of the other sciences." He can stand as a typical example of a thinker who refuses to acknowledge the hierarchy of Levels of Being and therefore cannot see any difference other than a difference in "complexity" between a stone and a man.

> The differences that do exist . . . [lie] chiefly in the greater or lesser complexity with which effects of the various laws are intertwined. . . .
> Another difference in scientific laws lies in the possibility of isolating their effects by experiment. . . . Certain sciences . . . can and do make extensive use of experiment. Certain others can use it but sparingly; others, such as the social sciences, little if any.[4]

With inanimate matter we can indeed experiment as we like; no amount of interference can destroy its life—for it has no life —or distort its inner experience—for there is no inner experience.

Experimentation is a valid and legitimate method of study only when it does not destroy the object under investigation. Inanimate matter cannot be destroyed; it can only be transformed. Life, consciousness, and self-awareness, on the other hand, are damaged very easily and almost invariably destroyed when the element of *freedom* inherent in these three powers is assumed to be nonexistent.

It is not simply the complexity at the higher Levels of Being which invalidates the use of the experimental method, but, much more importantly, the fact that causality, which rules supreme at the level of inanimate matter, is at the higher levels placed in a subservient position; it ceases to rule and is, instead, *employed* by higher powers for purposes unknown at the level of physics and chemistry.

When this point is missed and the attempt is made to press all sciences into the mold of physics, a certain kind of "progress" is indeed obtained; a kind of knowledge is accumulated which, however, more likely than not becomes a barrier to understanding and even a curse from which it is hard to escape. The lower takes the place of the higher, as when the study of a great work of art confines itself to the study of the materials of which it is made.

Physics, including chemistry and astronomy, is widely considered to be the most mature science and also the most successful. The life sciences, social sciences, and so-called humanities are thought to be less mature because they are beset by infinitely greater uncertainties. In terms of "maturity," we would have to say that the more mature the object of study, the less mature the science studying it. There is indeed more maturity in a human being than in a lump of mineral. That we have acquired more certain knowledge—of a kind—about the latter than about the former cannot surprise us if we remember that

if matter can be written   m
man has to be written $m + x + y + z$.

Physics deals only with "m", and it does so, as we have already seen, in a severely restrictive manner. Its program of investigation can be completed, just as the study of mechanics can be said to have been completed, and this may be called "maturity." It is certain that the study of "x," "y," and "z" can never be completed.

If we look carefully at what the various sciences in Field 4 actually do, we find that we can divide them roughly into two groups: those which are primarily *descriptive* of what can actually be seen or otherwise experienced and those which are primarily *instructional* of how certain systems work and can be made to produce predictable results. We might give botany as an example of the former, and chemistry of the latter. The difference between these two groups is seldom observed, with the result that most philosophies of science, in fact, are found to relate only to the instructional sciences and treat the descriptive ones as nonexisting. It is not, as has often been asserted, as if the difference between "descriptive" and "instructional" signified merely degrees of maturity or stages in the development of a science. F. S. C. Northrop claims that "Any empirical science in its normal healthy development begins with a more purely inductive emphasis . . . and then comes to maturity with deductively formulated theory *in which formal logic and mathematics play a most significant part.*"[5] This is perfectly true of "instructional" science—Northrop chooses geometry and physics as examples—instructional sciences par excellence —but it can never be true of descriptive sciences like botany, zoology, and geography, not to mention the historical sciences, whether they deal with nature or with man.

The distinction between descriptive and instructional sciences is similar to, but not identical with, that between "sciences for understanding" and "sciences for manipulation," which we discussed in an earlier chapter. A faithful *description* answers the question "What do I actually encounter?" An effective *instruction* answers the quite different question "What must I do to obtain a certain result?" Needless to say, neither descriptive sciences nor instructional sciences are mere accumulations of facts as presented by nature; in both cases, facts are "purified" or "idealized"; concepts are formed and theorems are put forward. A faithful description, however, is ruled by the concern "I must be careful not to leave out anything of significance," while an instruction is the more effective the more rigorously it excludes all factors that are not strictly necessary. People talk of "Okham's Razor," which is wielded to cut away everything that is superfluous to *obtaining results*. We can say therefore that descriptive science is—or should be— concerned primarily with *the whole truth*, while instructional science is concerned primarily only with such *parts or aspects of truth* as are useful for manipulation. In both cases I add the word "primarily" because this is not, and cannot be, a matter of an absolute difference.

Instructions, to be effective, must be precise, distinct, beyond doubt or dispute. It is not good enough to instruct: "Take a small quantity of water at a temperature that is comfortably warm." This may do for cooking but not for *exact* science. We must know precisely how much water and at precisely what temperature; there must be no room for "subjective" interpretation. Ideally, therefore, instructional science is totally quantified, and *qualities* (the color red, for instance) may play a part only when they "correlate" with some quantitatively definable phenomenon (such as light waves of a certain frequency). Its means of advance are logic and mathematics.

In the course of this advance it has been found that there is a strange and wonderful *mathematical* order in physical phenomena, and this has moved the minds of some of the most thoughtful modern physicists away from the crude materialism which ruled their science in the nineteenth century, and has made them aware of a transcendent reality. Even when traditional religion, which ascribed to God "the kingdom, the power, and the glory," remained unacceptable to them, they could not fail to recognize supreme mathematical talent somewhere in the construction and management of the Universe. Thus there has been, from the scientific side, a significant movement toward closing the infinitely harmful rift between natural science and religion. Some of the most advanced modern physicists would even agree with René Guénon's claim that "the whole of nature amounts to no more than a symbol of transcendent realities."[6]

If some physicists now think of God as a great mathematician, this is a significant reflection of the fact that "instructional science" *deals only with the dead aspect of nature.* Mathematics, after all, is far removed from life: At its heights it certainly manifests a severe kind of beauty and also a captivating elegance, which may even be taken as a sign of Truth; but, equally certainly, it has no warmth, none of life's messiness of growth and decay, hope and despair, joy and suffering. This must never be overlooked or forgotten: Physics and the other instructional sciences limit themselves to the lifeless aspect of reality, and this is necessarily so if the aim and purpose of science is to produce predictable results. Life, and, even more so, consciousness and self-awareness, cannot be ordered about; they have, we might say, a will of their own, which is a sign of maturity.

What we need to grasp at this point—and to inscribe on our *map of knowledge*—is this: Since physics and the other instructional sciences base themselves only on the dead aspect of nature, *they cannot lead to philosophy, if philosophy is to give us guidance on what* life *is all about.* Nineteenth-century physics told us that life was a cosmic accident, without meaning or purpose. The best twentieth-century physicists take it all back and tell us that they deal only with specific, strictly isolated systems, showing how these systems work, or can be made to work, and that no general philosophical conclusions can (or should) ever be drawn from this knowledge.

All the same, it is evident that the instructional sciences, even though they afford no guidance on how to conduct our lives, are *shaping our lives* through the technologies derived from them. Whether the results are for good or for evil is a question entirely outside their province. In this sense, it is correct to say that these sciences are ethically neutral. It remains true, however, that there is no science without scientists, and that questions of good and evil, even if they lie outside the province of science, cannot be considered to lie outside the province of the scientist. It is no exaggeration today to talk about a crisis of (instructional) science. If it continues to be a juggernaut outside humanistic control, there will be a reaction and revulsion against it, not excluding the possibility of violence.

Since the instructional sciences are concerned not with the whole truth but only with those parts or aspects of truth through which *results* can be obtained, it is proper that they should be judged exclusively by their results.

The claim that "Science" brings forth "Truth"—certain, unshakable, reliable knowledge which has been "scientifically proved"—and that this unique ability gives it a status higher than that of any other human activity—this claim on which the prestige of "Science" is founded needs to be investigated with some care. What is proof? We may hold a great many different theories. Can any of them be "proved"? We can see right away that it is possible to "prove" a recipe or any other instruction which takes the form of "If you do X, you will obtain Y." If such an instruction does not *work,* it is useless; if it does work, it has been "proved." *Pragmatism* is the philosophy which holds that the only valid test of truth is that *it works.* The pragmatist advises: "It is irrational to say: 'When an idea is true, it works'; you should say: 'When an idea works, it is true.'" In its purest form, however, pragmatism has the relative sterility of a hit-and-miss method. All sorts of instructions, taken in isolation, may be found to *work;* but unless I have some idea of a principle or "law" that makes a given system work, my chances of extending the range of instructional knowledge are slim.

The idea of proof, and therewith the idea of truth, in the instructional sciences is thus twofold: The instruction must work, i.e., lead to predicted results, and it must also be *intelligible* in terms of established scientific principles. Phenomena which are not intelligible in this sense are of no use, and therefore of no interest, to instructional science. It is a *methodological requirement* of the instructional sciences to ignore them. Such phenomena must not be allowed to call established scientific principles into question; there would be no pragmatic value in doing so.

Since the instructional sciences are concerned only with the amount of truth required to make their instructions effective and reliable, it follows that *proof* in these sciences suffers from the same limitations: It establishes that a certain set of instructions works and that there is *sufficient* truth in the underlying scientific principles to allow them to work, but it does not establish that other instructions might not also work or that an entirely different set of scientific principles might not also meet the case. As is well known, the pre-Copernican instructions on how to calculate the movements inside the solar system, based on the theory that the sun moved around the earth, for a long time produced much more accurate results than the post-Copernican instructions.

What, now, is the nature of *proof* in the descriptive sciences? The answer is inescapable: there can be classifications, observed regularities, speculations, theorems of different grades of plausibility, *but there can never be proof.* Scientific proof can exist only in instructional science, within the limitations mentioned above, because only that can be *proved* which we, with our minds or hands, can do ourselves. Our minds can *do* geometry, mathematics, and logic; we are therefore able to issue instructions *which work* and thereby establish *proof.* Equally, our hands are able to carry through a great variety of processes involving matter; we are therefore able to issue instructions on how to reach predetermined results and thereby establish proof. Without "doing" on the basis of instructions, there can be no proof.

As far as the instructional sciences are concerned, there can be no quarrel with pragmatism; on the contrary, this is precisely where pragmatism belongs, its proper place on the "map of knowledge." Nor can there be any quarrel with the restriction of the idea of truth to *intelligible* phenomena, which means disregarding unintelligible ones, and to theories of *heuristic* value, which means disregarding theories which prove "infertile" and fail to lead to an extension of instructional knowledge. These are the methodological requirements which, when rigorously observed, produce "progress," i.e., the enhancement of man's competence and power in employing natural processes for his own purposes.

Endless trouble, however, arises when the methodological requirements of the instructional sciences are taken as scientific methodology *per se*. Applied to the descriptive sciences, they lead to a methodology of error. The restrictions of pragmatism, heuristic principles, or Okham's razor are not compatible with truthful description. (The importance of this point will be further emphasized when we come to consider the Doctrine of Evolution.)

Physics and related instructional sciences deal with inanimate matter which, as far as we know, is devoid of life, consciousness, and self-awareness. At this Level of Being, there is nothing but "outer appearance," as distinct from "inner experience," and all we are concerned with is observable *facts*. Naturally, there can be nothing but *facts*, and when we say facts, we imply that they can be recognized by an observer. Unrecognized and—even more so—unrecognizable facts cannot and must not play any role in the theories of physics. It is therefore quite unproductive, at this level, to make a distinction between *what we can know* and *what actually exists*, i.e., between epistemology and ontology. When the modern physicist says: "In our experiments we sooner or later encounter ourselves," he is merely stating the obvious, namely, that the experimental results depend—not wholly but largely—on the question the physicist poses through his experimental arrangements. There is nothing mysterious about this, and it is quite wrong to conclude that it implies a disappearance of the difference between observer and observed. The Scholastic philosophers expressed this matter is a simple way: All knowledge is obtained *per modum cognoscentis*—in accordance with the cognitive powers of the knower.

The distinction between epistemology and ontology, or between *what we can know* and *what actually exists*, becomes significant only as we move higher up the Chain of Being. As an example, take the phenomenon of life. We can recognize the fact of life, and this recognition has led people to assert that "There exists in all living things an intrinsic factor—elusive, inestimable, and unmeasurable—that activates life."[7] So they talked about "vitalism." But this common-sense view is not acceptable to the instructional sciences. We are told that Ernest Nagel, a philosopher of science, rang "the death knell of vitalism" in 1951 by declaring vitalism a dead issue "because of the infertility of vitalism as a guide in biological research and because of the superior heuristic value of alternative approaches."[8]

The interesting and significant point is that this argument against vitalism is concerned not with its *truth* but with its *fertility*. To confuse these two is a very common error and causes a great deal of damage. A methodological principle—"fertility"—which is perfectly legitimate as a methodological principle, is substituted for the idea of *truth* and expanded into a philosophy with universal claims. As Karl Stern puts it, "methods become mentalities."[9] A statement is considered untrue, not because it appears to be incompatible with experience but because it does not serve as a guide in research and has no heuristic value; and, conversely, a theory is considered true, no matter how improbable it may be on general grounds, simply because of its "superior heuristic value."

The task of the descriptive sciences is to describe. The practitioners of these sciences know that the world is full of marvels which make all man's designs, theories, and other productions appear as a child's fumblings. This tends to induce in many of them an attitude of scientific humility. They are not attracted to their disciplines by the Cartesian idea of making themselves "masters and possessors of nature."[10] A faithful description, however, must be not only accurate but also *graspable* by the human mind, and endless accumulations of facts cannot be grasped; so there is an inescapable need for classifications, generalizations, explanations—in other words, for theories which offer some suggestion as to how the facts may "hang together." Such theories can never be "scientifically proved." The more comprehensive a theory in the descriptive sciences, the more is its acceptance an *act of faith*.

Comprehensive theories in the descriptive sciences can be divided into two groups: those which see *intelligence or meaning at work* in what they describe and those which see nothing but *chance and necessity*. It is obvious that neither the former nor the latter can be "seen," i.e., sensually experienced by man: In the Fourth Field of Knowledge there is only observation of movement and other kinds of material change; meaning or purpose, intelligence or chance, freedom or necessity, as well as life, consciousness, and self-awareness cannot be sensually observed. Only "signs" can be found and observed; the observer has to choose the *grade of significance* he is willing to attribute to them. To interpret them as signs of chance or necessity is as "unscientific" as to interpret them as signs of suprahuman intelligence; the one is as much an act of faith as the other. This does not mean that all interpretations on the *vertical scale*, signifying grades of significance or Levels of Being, are equally true or untrue; it means simply that their truth or untruth rests not on scientific proof but on right judgment, a power of the human mind which transcends mere logic just as the computer programmer's mind transcends the computer.

II

The distinctions which we are here discussing are of truly world-historical importance when we come to consider what is probably the most influential teaching of the modern age, the Evolutionist Doctrine. It is obvious that this doctrine cannot be classed with the instructional sciences: it belongs to the descriptive sciences. The question, therefore, is: "What does it describe?"

"Evolution in biology," says Julian Huxley, "is a loose and comprehensive term applied to cover any and every change occurring in the constitution of systematic units of animals and plants. . . ."[11] That there has been change in the constitution of species of animals and plants in the past is amply attested by the fossils found in the earth's crust; with the help of radioactive dating, they have been put into historical sequence with a very high degree of scientific certainty. Evolution, as a generalization within the descriptive science of biological change, can for this as well as for other reasons be taken as established beyond any doubt whatever.

The *Evolutionist Doctrine*, however, is a very different matter. Not content to confine itself to a systematic description of biological change, it purports to prove and explain it in much the same manner as proof and explanation are offered in the instructional sciences. This is a philosophical error with the most disastrous consequences.

"Darwin," we are told, "did two things: he showed that evolution was in fact contradicting scriptural legends of creation and that its cause, natural selection, was automatic with no room for divine guidance or design."[12] It should be obvious to anyone capable of philosophical thought that scientific observation as such can never do these "two things." "Creation," "divine guidance," and "divine design" are completely outside the possibility of scientific observation. Every animal or plant breeder knows beyond doubt that selection, including "natural selection," produces change; it is therefore scientifically correct to say that "natural selection has been *proved* to be an agent of evolutionary change." We can, in fact, prove it by doing. But it is totally illegitimate to claim that the discovery of this mechanism—natural selection—proves that evolution "was automatic with no room for divine guidance or design." It can be proved that people get money by finding it in the street, but no one would consider this sufficient reason for the assumption that all incomes are earned in this way.

The Doctrine of Evolutionism is generally presented in a manner which betrays and offends against all principles of scientific probity. It starts with the explanation of changes in living beings; then, without warning, it suddenly purports to explain not only the development of consciousness, self-awareness, language, and social institutions but also the origin of life itself. "Evolution," we are told, "is accepted by all biologists and natural selection is recognised as its cause." Since the origin of life is described as a "major step in evolution,"[13] we are asked to believe that inanimate matter is a masterful practitioner of natural selection. For the Doctrine of Evolutionism any possibility, no matter how remote, appears to be acceptable as if it were scientific proof that the thing actually happened:

> When a sample atmosphere of hydrogen, water vapour, ammonia, and methane was subjected to electric discharges and ultraviolet light, large numbers of organic compounds . . . were obtained by automatic synthesis. This proved that a prebiological synthesis of complex compounds was possible.[14]

On this basis we are expected to believe that living beings suddenly made their appearance by pure chance and, having done so, were able to maintain themselves in the general chaos:

> It is not unreasonable to suppose that life originated in a watery "soup" of prebiological organic compounds and that living organisms arose later by surrounding quantities of these compounds by membranes that made them into "cells." This is usually considered the starting point of organic ("Darwinian") evolution.[15]

One can just see it, can't one: organic compounds getting together and surrounding themselves by membranes—nothing could be simpler for these clever compounds—and lo! there is the cell, and once the cell has been born there is nothing to stop the emergence of Shakespeare, although it will obviously take a bit of time. There is therefore no need to speak of miracles *or to admit any lack of knowledge.* It is one of the great paradoxes of our age that people claiming the proud title of "scientist" dare to offer such undisciplined and reckless speculations as contributions to scientific knowledge, and *that they get away with it.*

Karl Stern, a psychiatrist with great insight, has commented thus:

> If we present, for the sake of argument, the theory of evolution in a most scientific formulation, we have to say something like this: "At a certain moment of time the temperature of the Earth was such that it became most favourable for the aggregation of carbon atoms and oxygen with the nitrogen-hydrogen combination, and that from random occurrences of large clusters molecules occurred which were most favourably structured for the coming about of life, and from that point it went on through vast stretches of time, until through processes of natural selection a being finally occurred which is capable of choosing love over hate and justice over injustice, of writing poetry like that of Dante, composing music like that of Mozart, and making drawings like those of Leonardo." Of course, such a view of cosmogenesis is crazy. And I do not at all mean crazy in the sense of slangy invective but rather in the technical meaning of psychotic. Indeed such a view has much in common with certain aspects of schizophrenic thinking.[16]

The fact remains, however, that this kind of thinking continues to be offered as objective science not only to biologists but to everybody eager to find out the truth about the origin, meaning, and purpose of human existence on Earth, and that, in particular, all over the world virtually all children are subjected to indoctrination along these lines.[17]

It is the task of science to observe and to report on its observations. It is not useful for it to *postulate* the existence of causative

agents, like a Creator, intelligences, or designers, who are outside all possibilities of direct observation. "Let us see how far we
can explain phenomena by observable causes" is an eminently
sensible and, in fact, very fruitful methodological principle.
Evolutionism, however, turns methodology into a faith which
excludes, *ex hypothesi,* the possibility of all higher grades of
significance. The whole of nature, which obviously includes
mankind, is taken as the product of chance and necessity *and
nothing else;* there is neither meaning nor purpose nor intelligence in it—"a tale told by an idiot, signifying nothing." This
is The Faith, and all contradicting observations have to be either ignored or interpreted in such a way that the The Faith is
upheld.

Evolutionism as currently presented has no basis in science.
It can be described as a peculiarly degraded religion, many of
whose high priests do not even believe in what they proclaim.
Despite widespread disbelief, the doctrinaire propaganda
which insists that the scientific knowledge of evolution *leaves
no room* for any higher faith continues unabated. Counterargu-
ments are simply ignored. The article on "Evolution" in *The
New Encyclopaedia Britannica* (1975) concludes with a section
entitled "The Acceptance of Evolution," which claims that "objections to evolution have come from theological and, for a
time, from political standpoints."[18] Who would suspect, reading
this, that the most serious objections have been raised by numerous biologists and other scientists of unimpeachable cre-
dentials? It is evidently thought unwise to mention them, and
books like Douglas Dewar's *The Transformist Illusion,*[19] which
offers an overwhelming refutation of Evolutionism on purely
scientific grounds, are not considered fit for inclusion in the
bibliography on the subject.

Evolutionism is not science; it is science fiction, even a kind
of hoax. It is a hoax that has succeeded too well and has impris-
oned modern man in what looks like an irreconcilable conflict
between "science" and "religion." It has destroyed all faiths
that pull mankind up and has substituted a faith that pulls mankind down. *"Nil admirari."* Chance and necessity and the utili-

tarian mechanism of natural selection may produce curiosities, improbabilities, atrocities, but nothing to be admired as an *achievement*—just as winning a prize in a lottery cannot elicit admiration. Nothing is "higher" or "lower"; everything is much of a muchness, even though some things are more complex than others—just by chance. Evolutionism, purporting to explain all and everything solely and exclusively by natural selection for adaptation and survival, is the most extreme product of the materialistic utilitarianism of the nineteenth century. The inability of twentieth-century thought to rid itself of this imposture is a failure which may well cause the collapse of Western civilization. For it is impossible for any civilization to survive without a faith in meanings and values transcending the utilitarianism of comfort and survival, in other words, without a religious faith.

"There can be little doubt," observes Martin Lings,

> that in the modern world more cases of loss of religious faith are to be traced to the theory of evolution as their immediate cause than to anything else. It is true, surprising as it may seem, that many people still contrive to live out their lives in a tepid and precarious combination of religion and evolutionism. But for the more logically minded, there is no option but to choose between the two, that is, between the doctrine of the fall of man and the "doctrine" of the rise of man, and to reject altogether the one not chosen. . . .
>
> Millions of our contemporaries have chosen evolutionism on the grounds that evolution is a "scientifically proved truth", as many of them were taught it at school; the gulf between them and religion is widened still further by the fact that the religious man, unless he happens to be a scientist, is unable to make a bridge between himself and them by producing the right *initial* argument, which must be on the scientific plane.[20]

If it is not on the "scientific plane," he will be shouted down "and reduced to silence by all sorts of scientific jargon." The truth of the matter, however, is that the initial argument must *not* be on the scientific plane; it must be *philosophical*. It amounts simply to this: that descriptive science becomes unscientific and illegitimate when it indulges in comprehensive explanatory theories which can be neither verified nor disproved by experiment. Such theories are not "science" but "faith."

III

What we can say at this stage of our exposition is that there is no possibility of deriving a *valid* FAITH from the study of the Fourth Field of Knowledge alone, which offers nothing but *observations of appearances.*

Still, it can be shown that the ever more precise, meticulous, conscientious, and imaginative observation of appearances, such as the best of modern scientists engage in, is producing an increasing amount of evidence which totally belies nineteenth-century materialistic utilitarianism. This is not the place for a detailed exposition of these findings. I can only mention again the conclusions reached by Wilder Penfield, which are supported, in a most interesting way, by the researches of Harold Saxton Burr, Professor Emeritus of Anatomy, Yale University School of Medicine. His "adventure in science" began in 1935 and continued for forty years: a search for the mysterious factor "x" which organizes inanimate material into living organisms and then maintains them. The molecules and cells of the human body are constantly disintegrating and new ones being rebuilt to replace them. "All protein in the body, for example, is 'turned over' every six months and in some organs such as the liver, the protein is renewed more frequently. When we meet a friend we have not seen for six months there is not one molecule in his face which was there when we last saw him."[21] Professor Burr and his collaborators discovered

that man—and, in fact, all *forms*—are ordered and controlled by electrodynamic fields which can be measured and mapped with precision. . . .

Though almost inconceivably complicated, the "fields of life" are of the same nature as the simpler fields known to modern physics and obedient to the same laws. Like the fields of physics, they are part of the organisation of the Universe and are influenced by the vast forces of space. Like the fields of physics, too, they have organising and directing qualities which have been revealed by many thousands of experiments.

Organisation and direction, the direct opposite of chance, imply purpose. So the fields of life offer purely electronic, instrumental evidence that man is no accident. On the contrary, he is an integral part of the Cosmos, embedded in its all-powerful fields, subject to its inflexible laws and a participant in the destiny and purpose of the Universe.[22]

The idea that the marvels of living nature are nothing but complex chemistry evolved through natural selection is thereby effectively destroyed, although the organizing power of *fields* remains a total mystery. Professor Burr's dethronement of chemistry, and therewith of biochemistry with all its DNA mythology of molecules becoming information systems, is certainly a very big step in the right direction. "To be sure," he says:

> chemistry is of great importance, because this is the gasoline that makes the buggy go, but the chemistry of a living system does not determine the functional properties of a living system any more than changing the gas makes a Rolls-Royce out of a Ford. The chemistry provides the energy, but the electrical phenomena of the electro-dynamic field determine the direction in which energy flows within the living system. Therefore they are of prime importance in understanding the growth and development of all living things.[23]

It is highly significant that as descriptive science becomes more refined and accurate, the rash utilitarian-materialistic doctrines of the nineteenth century are crumbling away one by one—in spite of the fact that most scientists insist on limiting their work to the Fourth Field of Knowledge, whereby, as we have shown, they methodically exclude all evidence of forces deriving from the higher Levels of Being and confine themselves to the dead aspect of the Universe. This methodical self-limitation makes sense—very good sense—for the instructional sciences, if only because the higher powers—life, consciousness, and self-awareness—are beyond "instruction": they *do* the instructing! But it makes no sense at all for the descriptive sciences. What is the value of a description if it omits the most interesting aspects and features of the object being described? Happily, there are now quite a number of scientists, like the zoologist Adolf Portmann and the botanist Heinrich Zoller (to name the two from whom I have benefited the most), who have had the courage to break out of the prison walls built by the modern Cartesians, and to show us the kingdom and the power and the glory of a mysteriously meaningful Universe.

IV

The Four Fields of Knowledge can be clearly distinguished; nevertheless, knowledge itself is a unity. The main purpose of showing the four fields separately is to make the unity appear in its plenitude. A few examples may be given of what this analysis helps us to understand:

1. The unity of knowledge is destroyed when one or several of the Four Fields of Knowledge remain uncultivated, and also when a field is cultivated with instruments and methodologies which are appropriate only in quite another field.

2. To obtain a clear view of Reality it is necessary to relate the Four Fields of Knowledge to the four Levels of Being. We have already touched on this in stating that little can be learned about human nature by anyone who confines his studies to the Fourth Field of Knowledge, the field of appearances. Similarly, little if anything can normally be learned about the mineral kingdom from studies of one's own inner experiences, unless certain higher sensitivities have been developed, as in the cases of people like Lorber, Cayce, and Therese Neumann.

3. The instructional sciences do well to confine their attention exclusively to Field 4, since only in this field of appearances can mathematical precision be obtained. The descriptive sciences, on the other hand, betray their calling when they ape the instructional sciences and confine themselves to the obser-

vation of appearances. If they cannot penetrate to *meaning* and *purpose*—ideas derivable only from inner experience (Fields 1 and 2)—they remain sterile and can be of use to humanity only as producers of "inventories," which hardly deserves the noble name of science.

4. Self-knowledge, so universally praised as the most valuable, remains worse than useless if it is based solely on the study of Field 1, one's own inner experiences; it *must* be balanced by an equally intensive study of Field 3, through which we learn to know ourselves as others know us. This point is all too often overlooked through failure to distinguish between Field 1 and Field 3.

5. Finally, there is social knowledge, that is, the knowledge needed for the establishing of harmonious relationships among people. Since we have no direct access to Field 2—the inner experiences of other beings—obtaining indirect access is one of man's most important tasks as a social being. This indirect access can be gained only through self-knowledge, which shows that it is a grave error to accuse a man who pursues self-knowledge of "turning his back on society." The opposite would be more nearly true: that a man who fails to pursue self-knowledge is and remains a danger to society, for he will tend to misunderstand everything that other people say or do, and remain blissfully unaware of the significance of many of the things he does himself.

10

Two Types of Problems

First, we dealt with "The World"—its four Levels of Being; second, with "Man"—his equipment for meeting the world: to what extent is it adequate for the encounter? And third, we dealt with learning about the world and about oneself—the "Four Fields of Knowledge." It remains to examine what it means to live in this world.

To live means to *cope*, to contend and keep level with all sorts of circumstances, many of them difficult. Difficult circumstances present *problems*, and it might be said that living means, above all else, dealing with problems.

Unsolved problems tend to cause a kind of existential anguish. Whether this has always been so may well be questioned; but it is certainly so in the modern world, and one of the weapons in the modern battle against anguish is the Cartesian approach: "Deal only with ideas that are distinct, precise, and certain beyond any reasonable doubt; therefore, rely on geometry, mathematics, quantification, measurement, and exact observation." This is the way, the only way (we are told) to solve problems; this is the road, the only road, of progress; if only we abandon all sentiment and other irrationalities, all problems can and will be solved. We live in the age of the Reign of

Quantity.[1] Quantification and cost/benefit analysis are said to be the answer to most, if not all, of our problems, although where we are dealing with somewhat complex beings, like humans, or complex systems, like societies, it may still take a bit of time until sufficient data have been assembled and analyzed. Our civilization is uniquely expert in problem-solving. There are more scientists and people applying the "scientific" method at work in the world today than there have been in all previous generations added together, and they are not wasting their time contemplating the marvels of the Universe or trying to acquire self-knowledge: they are *solving problems*. (I can imagine someone becoming slightly anxious at this point and inquiring: "If this is so, aren't we running out of problems?" It would be easy to reassure him: We have more and bigger problems now than any previous generation could boast, including problems of survival.)

This extraordinary situation might lead us to inquire into the nature of *"problems."* We know there are *solved* problems and *unsolved* problems. The former, we may feel, present no issue; but as regards the latter: Are there not problems that are not merely unsolved but insoluble?

First, let us look at solved problems. Take a design problem —say, how to make a two-wheeled, man-powered means of transportation. Various solutions are offered which gradually and increasingly *converge* until, finally, a design emerges which is "the answer"—a bicycle—an answer that turns out to be amazingly stable over time. Why is this answer so stable? Simply because it complies with the laws of the Universe—laws at the level of inanimate nature.

I propose to call problems of this nature *convergent* problems. The more intelligently you (whoever you are) study them, the more the answers *converge*. They may be divided into "convergent problem *solved"* and "convergent problem *as yet unsolved."* The words "as yet" are important, for there is no reason *in principle* why they should not be solved some day. Everything takes time, and there simply has not yet been time enough to get around to solving them. What is needed is more time, more money for research and development (R & D) and, maybe, more talent.

It also happens, however, that a number of highly able people may set out to study a problem and come up with answers which contradict one another. They do *not converge*. On the contrary, the more they are clarified and logically developed, the more they *diverge*, until some of them appear to be the exact *opposites* of the others. For example, life presents us with a very big problem—not the technical problem of two-wheeled transport, but the human problem of how to educate our children. We cannot escape it; we have to face it, and we ask a number of equally intelligent people to advise us. Some of them, on the basis of a clear intuition, tell us: "Education is the process by which existing culture is passed on from one generation to the next. Those who have (or are presumed to have) knowledge and experience *teach,* and those who as yet lack knowledge and experience *learn.* For this process to be effective, authority and discipline must be set up." Nothing could be simpler, truer, more logical and straightforward. Education calls for the establishment of *authority* for the teachers and *discipline* and *obedience* on the part of the pupils.

Now, another group of our advisers, having gone into the problem with the utmost care, says this: "Education is nothing more nor less than the provision of a *facility.* The educator is like a good gardener, whose function is to make available healthy, fertile soil in which a young plant can grow strong roots; through these it will extract the nutrients it requires. The young plant will develop in accordance with its own laws of being, which are far more subtle than any human can fathom, and will develop best when it has the greatest possible freedom to choose exactly the nutrients it needs." In other words, education as seen by this second group calls for the establishment, not of discipline and obedience, but of freedom—the greatest possible freedom.

If our first group of advisers is right, discipline and obedience are "a good thing," and it can be argued with perfect logic that if something is "a good thing," more of it would be a better thing, and perfect discipline and obedience would be a perfect thing . . . and the school would become a prison house.

Our second group of advisers, on the other hand, argues that in education freedom is "a good thing." If so, more freedom would be an even better thing, and perfect freedom would produce perfect education. The school would become a jungle, even a kind of lunatic asylum.

Freedom and discipline (obedience) here is a pair of perfect opposites. No compromise is possible. It is either the one or the other. It is either "Do as you like" or "Do as I tell you."

Logic does not help us because it insists that if a thing is true its opposite cannot be true at the same time. It also insists that if a thing is good, more of it will be better. Here we have a very typical and very basic problem, which I call a *divergent problem,* and it does not yield to ordinary, "straight-line" logic; it demonstrates that *life is bigger than logic.*

"What is the best method of education?" presents, in short, a divergent problem par excellence. The answers tend to diverge, and the more logical and consistent they are, the greater is the divergence. There is "freedom" *versus* "discipline and obedience." There is no solution. And yet some educators are better than others. How does this come about? One way to find out is to ask them. If we explained to them our philosophical difficulties, they might show signs of irritation with this intellectual approach. "Look here," they might say, "all this is far too clever for me. The point is: You must *love* the little horrors." Love, empathy, *participation mystique,* understanding, compassion—these are faculties of a *higher order* than those required for the implementation of any policy of discipline or of freedom. To mobilize these higher faculties or forces, to have them available not simply as occasional impulses but permanently, requires a high level of self-awareness, and that is what makes a great educator.

Education presents the classical example of a divergent problem, and so of course does politics, where the most frequently encountered pair of opposites is "freedom" and "equality," which in fact means freedom *versus* equality. For if natural forces are left free, i.e., left to themselves, the strong will prosper and the weak will suffer, and there will be no trace of equality. The enforcement of equality, on the other hand, requires the curtailment of freedom—*unless something intervenes from a higher level*.

I do not know who coined the slogan of the French Revolution*; he must have been a person of rare insight. To the pair of opposites, *Liberté* and *Egalité*, irreconcilable in ordinary logic, he added a third factor or force—*Fraternité*, brotherliness—which comes from a higher level. How do we recognize this fact? Liberty or equality can be instituted by legislative action backed by force, but brotherliness is a human quality beyond the reach of institutions, beyond the level of manipulation. It can be achieved only by individual persons mobilizing their own higher forces and faculties, in short, becoming better people. "How do you make people become better?" That this is a question constantly being asked merely shows that the essential point is being missed altogether. *Making* people better belongs to the level of manipulation, the same level at which the opposites exist and where their reconciliation is impossible.

The moment we recognize that there are two different *types* of problems with which we have to deal on our journey through life—"convergent" and "divergent" problems—some very interesting questions arise in our minds:

> How can I recognize whether a problem belongs to the one type or to the other?
> What constitutes the difference?
> What constitutes the solution of a problem in each of the two types?
> Is there "progress"? Can solutions be accumulated?

The attempt to deal with questions of this kind will undoubtedly lead to many further explorations.

Let us begin then with the question of recognition. With a convergent problem, as we said, the answers suggested for its solution tend to converge, to become increasingly precise, until

---

*Some people say it was Louis-Claude de Saint-Martin (1743–1803) who signed his works Le Philosophe inconnu, the Unknown Philosopher.

finally they can be written down in the form of an instruction. Once the answer has been found, the problem ceases to be interesting: A solved problem is a dead problem. To make use of the solution does not require any higher faculties or abilities —the challenge is gone, the work is done. Whoever makes use of the solution can remain relatively passive; he is a recipient, getting something for nothing, as it were. Convergent problems relate to the *dead* aspect of the Universe, where manipulation can proceed without let or hindrance and where man can make himself "master and possessor," because the subtle, higher forces—which we have labeled life, consciousness, and self-awareness—are not present to complicate matters. Wherever these higher forces intervene to a significant extent, the problem ceases to be convergent. We can say, therefore, that *convergence* may be expected with regard to any problem which does not involve life, consciousness, self-awareness, which means in the fields of physics, chemistry, astronomy, and also in abstract spheres like geometry and mathematics, or games like chess.

The moment we deal with problems involving the higher Levels of Being, we must expect *divergence*, for there enters, to however modest a degree, the element of freedom and inner experience. In them we can see the most universal pair of opposites, the very hallmark of Life: growth and decay. Growth thrives on freedom (I mean healthy growth; pathological growth is really a form of decay), while the forces of decay and dissolution can be contained only through some kind of order. These basic pairs of opposites

Growth *versus* Decay
and Freedom *versus* Order

are encountered wherever there is life, consciousness, self-awareness. As we have seen, it is pairs of opposites that make a problem divergent, while the absence of pairs of opposites (of this basic character) ensures convergence.

The methodology of problem-solving, as can easily be observed, is what we might call "the laboratory approach." It

consists of eliminating all factors which cannot be strictly controlled or, at least, accurately measured and "allowed for."
What remains is no longer a part of real life, with all its un-
predictabilities, but an isolated system posing convergent, and
therefore in principle *soluble*, problems. At the same time, the
solution of a convergent problem *proves* something about the
isolated system, but nothing at all about matters outside and
beyond it.

I have said that to solve a problem is to kill it. There is nothing
wrong with "killing" a convergent problem, for it relates to
what remains after life, consciousness, and self-awareness have
already been eliminated. But can—or should—divergent problems be killed? (The words "final solution" still have a terrible
ring in the ears of my generation.)

Divergent problems cannot be killed; they cannot be solved
in the sense of establishing a "correct formula"; they can, however, be transcended. A pair of opposites—like freedom and
order—are opposites at the level of ordinary life, but they cease
to be opposites at the higher level, the really *human* level,
where self-awareness plays its proper role. It is then that such
higher forces as love and compassion, understanding and empathy, become available, not simply as occasional impulses (which
they are at the lower level) but as a regular and reliable resource. Opposites cease to be opposites; they lie down together
peacefully like the lion and the lamb in Dürer's famous picture
of Saint Hieronymus (who himself represents "the higher
level").

How can opposites cease to be opposites when a "higher
force" is present? How is it that liberty and equality cease to be
mutually antagonistic and become "reconciled" when brotherliness is present? These are not logical but *existential* questions.
The main concern of existentialism, it has been noted,[2] is that
experience has to be admitted as evidence, which implies that
without experience there is no evidence. That opposites are
transcended when "higher forces"—like love and compassion
—intervene is not a matter to be argued in terms of logic: it has
to be experienced in one's actual existence (hence: "existential-

ism"). Here is a family, let us say, with two big boys and two small girls; freedom prevails, and it does not destroy equality because *brotherliness* controls the use of the superior power possessed by the big boys.

It is important for us to become fully aware of these pairs of opposites. Our logical mind does not like them: it generally operates on the either/or or yes/no principle, like a computer. So, at any time it wishes to give its exclusive allegiance to either one or the other of the pair, and since this exclusiveness inevitably leads to an ever more obvious loss of realism and truth, the mind may suddenly change sides, often without even noticing it. It swings like a pendulum from one opposite to the other, and each time there is a feeling of "making up one's mind afresh"; or the mind may become rigid and lifeless, fixing itself on one side of the pair of opposites and feeling that now "the problem has been solved."

The pairs of opposites, of which *freedom and order* and *growth and decay* are the most basic, put tension into the world, a tension that sharpens man's sensitivity and increases his self-awareness. No real understanding is possible without awareness of these pairs of opposites which permeate everything man does.

In the life of societies there is the need for both justice and mercy. "Justice without mercy," said Thomas Aquinas, "is cruelty; mercy without justice is the mother of dissolution"[3]—a very clear identification of a divergent problem. Justice is a denial of mercy, and mercy is a denial of justice. Only a higher force can reconcile these opposites: wisdom. The problem cannot be solved, but wisdom can transcend it. Similarly, societies need stability *and* change, tradition *and* innovation, public interest *and* private interest, planning *and* laissez-faire, order *and* freedom, growth *and* decay. Everywhere society's health depends on the simultaneous pursuit of mutually opposed activities or aims. The adoption of a final solution means a kind of death sentence for man's humanity and spells either cruelty or dissolution, generally both.

Divergent problems offend the logical mind, which wishes to

remove tension by coming down on one side or the other, but they provoke, stimulate, and sharpen the higher human faculties, without which man is nothing but a clever animal. A refusal to accept the divergency of divergent problems causes these higher faculties to remain dormant and to wither away, and when this happens, the "clever animal" is more likely than not to destroy itself.

Man's life can thus be seen and understood as a succession of divergent problems which must inevitably be encountered and have to be coped with in some way. They are refractory to mere logic and discursive reason, and constitute, so to speak a strain-and-stretch apparatus to develop the Whole Man, and that means to develop man's supralogical faculties. All traditional cultures have seen life as a school and have recognized, in one way or another, the essentiality of this teaching force.

II

At this point, it may be appropriate to say a few words about art. Today, as far as art is concerned, there seems to be nothing at all to go by and anything will do. Who dares to say "boo" to anything claiming to be "art ahead of its time"? However, we need not be so timid. We can obtain reliable bearings by relating art to the human being, which consists of feeling, thinking, and willing. If art aims primarily to affect our feelings, we may call it entertainment; if it aims primarily to affect our will, we may call it propaganda. These two, entertainment and propaganda, we can recognize as a pair of opposites, and we have no difficulty in sensing that something is missing. No great artist has ever turned his back on either entertainment or propaganda, nor was he ever satisfied with just these two. Invariably he strove to communicate truth, the *power* of truth, by appealing to man's higher intellectual faculties, which are suprarational. Entertainment and propaganda by themselves do not give us power but exert power over us. When they are transcended by, and made subservient to, the communication of

Truth, art helps us to develop our higher faculties, and this is what matters.

If art is to have any real value, says Ananda K. Coomaraswamy,

> if it is to nourish and make the best part of us grow, as plants are nourished and grow in suitable soils, it is to the understanding and not to fine feelings that an appeal must be made. In one respect the public is right; it always wants to know what a work of art is "about." ... Let us tell them the painful truth that most of these [great] works of art are about God, whom we never mention in polite society. Let us admit that if we are to offer an education in agreement with the innermost nature and eloquence of [these great works of art] themselves, that this will not be an education in sensibility, but an education in philosophy, in Plato's and Aristotle's sense of the word, for whom *it means ontology and theology and the map of life, and a wisdom to be applied to everyday matters.*[4]

All great works of art are "about God" in the sense that they show the perplexed human being the path, the way up the mountain, providing a Guide for the Perplexed. We may again remind ourselves of one of the greatest examples of such art, Dante's *Divine Comedy*. Dante wrote for ordinary men and women, not for people with sufficient private means to be interested mainly in fine feelings. "The whole work," he explains, "was undertaken not for a speculative but a practical end ... the purpose of the whole is to remove those living in this life from a state of misery, and lead them into a state of felicity."[5] The pilgrim—Dante himself—*nel mezzo del cammin di nostra vita,* that is, at the height of his powers and outward success, suddenly realizes that he is not at the height at all but, on the contrary, "in a dark wood, where the right way was lost."

> Ah! how hard a thing it is to tell
> what this wild and rough and difficult wood was,
> which in thought renews my fear!
> So bitter that death is little more.

He cannot remember how he ever got there,

so full was I of slumber at that moment
when I abandoned the true way.

Having "found himself," Dante looks up and sees the mountain,

clothed already with the rays of the planet [the sun]
which leads man aright along every path,

the very mountain he had meant to climb. He makes a new
attempt, but he finds his way barred by three animals: first,

at the beginning of the steep
a she-leopard, light and very nimble,
which was covered with a spotted coat.
And she did not withdraw from before my face,
nay, hindered so my road *that I often turned*
        to go back.

Light, very nimble, with a spotted coat—all the pleasant temptations of life, to which he was used to yielding. There is worse
to come: a lion, fearful in his pride, and a she wolf

which in her leanness seemed laden with all cravings,
and ere now had made many folk to live forlorn,—
she brought on me so much heaviness,
with the fear that came from sight of her,
that I lost hope. . . .

Dante, however, is seen "from heaven" by Beatrice, who
wants to help him. She cannot do so herself, as he has sunk too
low for religion to reach him, and so she asks *Art*, in the person
of Virgil, to guide him out of "this savage place." True art is the
intermediary between man's ordinary nature and his higher
potentialities, and so Dante accepts Virgil:

"Thou by thy words has so disposed my heart
with desire of going,
that I have returned to my first intent.
Now go, for one sole will is in us both:
thou leader, thou lord, and thou master."[6]

Only the truth can be accepted as leader, lord, and master. To
treasure art simply for its beauty is to miss the point. The true

function of art is "so to dispose [the] heart with desire of going" "up the mountain," *which is what we really wish to do but keep forgetting,* that we "return to our first intent."

The whole of great literature deals with divergent problems. To read such literature—even the Bible!—simply "as literature," as if its main purpose were poetry, imagination, artistic expression with an especially apt use of words and similes, is to turn the sublime into the trivial.

III

Many people today call for a new moral basis of society, a new foundation of ethics. When they say "new," they seem to forget that they are dealing with divergent problems, which call not for new inventions but for the development of man's higher faculties and their application. "Some rise by sin, and some by virtue fall," says Shakespeare in *Measure for Measure,* suggesting that it is not good enough to decide that virtue is good and vice is bad (which they are!), but that the important thing is whether a person *rises* to his higher potentialities or *falls* away from them. Normally, men rise through virtue, but if virtue is merely external and lacks inner power, it makes them merely complacent and they fail to develop. Similarly, what by ordinary standards is sin may set in motion the all-important process of development if its shock causes a man to awaken his higher faculties which have previously been asleep. To quote an example from the Eastern traditions: "By what men fall by that they rise," says the Kular nava Tantra.

All traditional wisdom, of which both Dante and Shakespeare are outstanding representatives, transcends ordinary, calculating logic and defines "The Good" as that which helps us to become truly human by developing our higher faculties— which are conditional on, and also part of, self-awareness. Without them there is no humanity, as distinct from the animal kingdom, and the question of what is "The Good" reduces itself to Darwinian questions of adaptation and survival and the utilitarianism of "the greatest happiness of the greatest number," where happiness rarely implies anything more than comfort and excitement.

In fact, however, people do not accept these "reductions." Even when, being well adapted, they survive with plenty of comfort and excitement, they go on asking: "What is 'Good'? What is 'Goodness'? What is 'Evil'? What is 'Sin'? What must I do to live a worthwhile life?"

In the whole of philosophy, there is no subject in greater disarray than ethics. Anyone asking the professors of ethics for the bread of guidance or how to conduct himself, will receive not even a stone but just a torrent of "opinions." With very few exceptions, they embark upon an investigation into ethics without any prior clarification of the purpose of human life on Earth. It is obviously impossible to decide what is good or bad, right or wrong, virtuous or evil, without an idea of purpose: Good for what? To raise the question of purpose has been called "the naturalistic fallacy"—virtue is its own reward! None of the great teachers of mankind would have been satisfied with such an evasion. If a thing is said to be good but no one can tell me what it is good *for,* how can I be expected to take any interest in it? If our guide, our annotated Map of Life, cannot show us where The Good is situated and how it can be reached, it is worthless.

Let us recapitulate. The first Great Truth we have discussed is the hierarchic structure of the World: at least four great Levels of Being, with new powers added as we move up the Chain of Being. At the human level, we can clearly perceive that it is open-ended. There is no discernible limit to what Man can do; he seems to be *"capax universi,"* as the Ancients used to say, and what *one* person has done shines thereafter like a light in darkness as a capability of Man, even if no second person is ever found able to do it again. The human being, even in full maturity, is obviously not a finished product, although some are undoubtedly more "finished" than others. With most people, the specifically human faculty of self-awareness remains, until the end of their lives, only the germ of a faculty, so underdeveloped that it rarely becomes active, and then only for brief moments. This is precisely the "talent" which according to traditional teachings we can and should develop threefold, even tenfold, and which we should on no account bury in the ground for safekeeping.

We have been able to touch only lightly on the various "progressions" we notice when contemplating the four Levels of Being: from lifeless mineral to the self-aware person and onward—to the most perfect, most thoroughly integrated, enlightened, free "Person" we can conceive. These extrapolations help us not only to obtain a clearer understanding of what our ancestors were concerned with when they talked about God but also to recognize the one and only direction our life on Earth must *develop* if it is to have sense and meaning.

The second Great Truth is that of *adaequatio*—that everything in the world around us must be matched, as it were, with some sense, faculty, or power within us; otherwise we remain unaware of its existence. There is, therefore, a hierarchic structure of gifts inside us, and, not surprisingly, the higher the gift, the more rarely is it to be found in a highly developed form, and the greater are the efforts required for its development. To enhance our Level of Being, we have to adopt a life-style conducive to such enhancement, which means one that will grant our lower nature no more attention and care than it requires and will leave us with ample free time and attention to pursue our higher development.

A central part of this pursuit is the cultivation of the Four Fields of Knowledge. The quality of our understanding depends decisively on the detachment, objectivity, and care with which we learn to study ourselves—both what goes on inside us (Field 1) and how we appear as objective phenomena in the eyes of others (Field 3). Instruction on cultivating self-knowledge of this dual kind is the main content of all traditional religious teachings but has been almost entirely lacking in the West for the last hundred years. That is why we cannot trust one another, why most people live in a state of continuous anxiety, why despite all our technologies communication becomes ever more difficult, and why we need ever more organized *welfare* to plaster over the gaping holes torn by the progressive disappearance of spontaneous social cohesion. The Christian (and other) saints knew themselves so well that they could "see into" other beings. The idea that Saint Francis could communicate with animals, birds, even flowers, must of course seem incredible to modern men who have so neglected self-knowledge that they have difficulties communicating even with their wives.

The "inner world," seen as fields of knowledge (Field 1 and Field 2), is the world of freedom; the "outer world" (Field 3 and Field 4) is the world of necessity. All our serious problems of living are suspended, as it were, between these two poles of freedom and necessity. They are *divergent* problems, not for solving. Our anxiety to *solve* problems stems from our lack of self-knowledge, which has created the kind of existential anguish of which Kierkegaard is one of the early and most impressive exponents. The same anxiety to *solve* problems has led to a virtually total concentration of intellectual effort on the study of *convergent* problems.

Great pride is taken in this voluntary limitation of the limitless Intellect to "the art of the soluble." "Good scientists," says P. B. Medawar, "study the most important problems they think they can solve. It is, after all, their professional business to solve problems, not merely to grapple with them."[7] This is fair enough; it clearly demonstrates, at the same time, that "good scientists" in this sense can deal only with the dead aspect of the Universe. But the real problems of life have to be *grappled with*. To repeat the quotation from Thomas Aquinas, "The slenderest knowledge that may be obtained of the highest things is more desirable than the most certain knowledge obtained of lesser things," and "grappling" with the help of slender knowledge is the real stuff of life, whereas solving problems (which, to be soluble, must be convergent) with the help of "the most certain knowledge obtained of lesser things" is merely one of many useful and perfectly honorable human activities designed to save labor.

While the logical mind abhors divergent problems and tries to run away from them, the higher faculties of man accept the challenges of life as they are offered, without complaint, knowing that when things are most contradictory, absurd, difficult, and frustrating, then, *just then*, life really makes sense: as a mechanism provoking and almost forcing us to develop toward higher Levels of Being. The question is one of faith, of choosing our own "grade of significance." Our ordinary mind always tries to persuade us that we are nothing but acorns and that our greatest happiness will be to become bigger, fatter, shinier acorns; but that is of interest only to pigs. Our faith gives us knowledge of something much better: that we can become oak trees.

What is good and what is bad? What is virtuous and what is evil? It all depends on our faith. Taking our bearings from the four Great Truths discussed in this book and studying the interconnections between these four landmarks on our "map," we do not find it difficult to discern what constitutes the true progress of a human being:

1. One's first task is to learn from society and "tradition" and to find one's temporary happiness in receiving directions from outside.
2. One's second task is to interiorize the knowledge one has gained, sift it, sort it out, keeping the good and jettisoning the bad; this process may be called "individuation," becoming self-directed.
3. One's third task cannot be tackled until one has accomplished the first two, and is one for which one needs the very best help that can possibly be found: It is "dying to oneself," to one's likes and dislikes, to all one's egocentric preoccupations. To the extent that one succeeds in this, one ceases to be directed from outside, and also ceases to be self-directed. One has gained freedom or, one might say, one is then God-directed. If one is a Christian, that is precisely what one would hope to be able to say.

If this is the threefold task before each human being, we can say that "good" is what helps me and others along on this journey of liberation. I am called upon to "love my neighbor as myself," but I cannot love him at all (except sensually or sentimentally) unless I have loved myself sufficiently to embark on the journey of development as described. How can I love and help him as long as I have to say, with Saint Paul: "My own behavior baffles me. For I find myself not doing what I really want to do but doing what I really loathe"? In order to become capable of loving and helping my neighbor as well as myself, I am called upon to "love God," that is, strenuously and patiently to keep my mind straining and stretching toward the highest things, to Levels of Being above my own. Only there lies "goodness" for me.

Epilogue

After Dante (in the *Divine Comedy*) had "woken up" and found himself in the horrible dark wood where he had never meant to go, his good intention to make the ascent up the mountain was of no avail; he first had to descend into the *Inferno* to be able fully to appreciate the reality of sinfulness. Today, people who acknowledge the *Inferno of things as they really are* in the modern world are regularly denounced as "doomwatchers," pessimists, and the like. Dorothy Sayers, one of the finest commentators on Dante as well as on modern society, has this to say:

> That the *Inferno* is a picture of human society in a state of sin and corruption, everybody will readily agree. And since we are today fairly well convinced that society is in a bad way and not necessarily evolving in the direction of perfectibility, we find it easy enough to recognise the various stages by which the deep of corruption is reached. Futility; lack of a living faith; the drift into loose morality, greedy consumption, financial irresponsibility, and uncontrolled bad temper; a self-opinionated and obstinate individualism; violence, sterility, and lack of reverence for life and property including one's own; the exploitation of sex, the debasing of language by advertisement and propaganda, the commercialising of religion, the pandering to superstition and the conditioning of people's minds by mass-hysteria and "spell-binding" of all kinds, venality and string-

pulling in public affairs, hypocrisy, dishonesty in material things, intellectual dishonesty, the fomenting of discord (class against class, nation against nation) for what one can get out of it, the falsification and destruction of all the means of communication; the exploitation of the lowest and stupidest mass-emotions; treachery even to the fundamentals of kinship, country, the chosen friend, and the sworn allegiance: these are the all-too-recognisable stages that lead to the cold death of society and the extinguishing of all civilised relations.[1]

What an array of divergent problems! Yet people go on clamoring for "solutions" and become angry when they are told that the restoration of society must come from within and cannot come from without. The above passage was written a quarter of a century ago. Since then, there has been further progress downhill, and the description of the *Inferno* sounds even more familiar.

But there have also been positive changes: Some people are no longer angry when told that *restoration must come from within;* the belief that everything is "politics" and that radical rearrangements of the "system" will suffice to save civilization is no longer held with the same fanaticism as it was held twenty-five years ago. Everywhere in the modern world there are experiments in new life-styles and Voluntary Simplicity; the arrogance of materialistic Scientism is in decline, and it is sometimes tolerated even in polite society to mention God.

Admittedly, some of this change of mind stems initially not from spiritual insight but from materialistic fear aroused by the environmental crisis, the fuel crisis, the threat of a food crisis, and the indications of a coming health crisis. In the face of these —and many other—threats, most people still try to believe in the "technological fix." If we could develop *fusion energy,* they say, our fuel problems would be solved; if we would perfect the processes of turning oil into edible proteins, the world's food problem would be solved; and the development of new drugs will surely avert any threat of a health crisis . . . and so on.

All the same, faith in modern man's omnipotence is wearing thin. Even if all the "new" problems were solved by technological fixes, the state of futility, disorder, and corruption would remain. It existed before the present crises became acute, and it will not go away by itself. More and more people are beginning to realize *that "the modern experiment" has failed.* It received its early impetus from what I have called the Cartesian revolution, which, with implacable logic, separated man from those Higher Levels that alone can maintain his humanity. Man closed the gates of Heaven against himself and tried, with immense energy and ingenuity, to confine himself to the Earth. He is now discovering that the Earth is but a transitory state, so that a refusal to reach for Heaven means an involuntary descent into Hell.

It may conceivably be possible to live without churches; but it is not possible to live without religion, that is, without systematic work to keep in contact with, and develop toward, Higher Levels than those of "ordinary life" with all its pleasure or pain, sensation, gratification, refinement or crudity—whatever it may be. *The modern experiment to live without religion has failed,* and once we have understood this, we know what our "post modern" tasks really are. Significantly, a large number of young people (of varying ages!) are looking in the right direction. They feel in their bones that the ever more successful solution of convergent problems is of no help at all—it may even be a hindrance—in learning how to cope, to grapple, with the divergent problems which are the stuff of real life.

The art of living is always to make a good thing out of a bad thing. Only if we *know* that we have actually descended into *infernal regions* where nothing awaits us but "the cold death of society and the extinguishing of all civilised relations," can we summon the courage and imagination needed for a "turning around," a *metanoia.* This then leads to seeing the world in a new light, namely, as a place where the things modern man continuously talks about and always fails to accomplish *can actually be done.* The generosity of the Earth allows us to feed all mankind; we know enough about ecology to keep the Earth a healthy place; there is enough room on the Earth, and there are enough materials, so that everybody can have adequate shelter; we are quite competent enough to produce sufficient supplies of necessities so that no one need live in misery. Above all, we shall then see that the economic problem is a convergent problem *which has been solved already:* we know how to provide *enough* and do not require any violent, inhuman, aggressive technologies to do so. There *is* no economic problem and, in a sense, there never has been. But there is a moral problem, and moral problems are not convergent, capable of being solved so that future generations can live without effort. No, they are divergent problems, which have to be understood and transcended.

Can we rely on it that a "turning around" will be accomplished by enough people quickly enough to save the modern world? This question is often asked, but no matter what the answer, it will mislead. The answer "Yes" would lead to complacency, the answer "No" to despair. It is desirable to leave these perplexities behind us and get down to work.

Notes

Chapter 1. On Philosophical Maps

1. To be precise: in August, 1968, during the week of the Soviet invasion of Czechoslovakia.

2. *Summa theologica* I,1,5 ad 1.

3. Maurice Nicoll, *Psychological Commentaries*, Vol. 1 (London, 1952).

4. A. Koestler & J. R. Smythies (eds.), *Beyond Reductionism*. Chapter on "Reductionism and Nihilism" by Victor E. Frankl (London, 1969).

5. *Ibid.*

6. Quoted by Michael Polanyi, *Personal Knowledge* (London, 1958).

7. Koestler and Smythies, *op. cit.*

8. Plato, *Symposium*. Jowett translation (Oxford, 1871).

9. Cf. F. S. C. Northrop, *The Logic of the Science* and *Humanities* (New York, 1959).

10. W. Y. Evans-Wentz, *Tibetan Yoga and Secret Doctrines* (Oxford, 1935).

11. René Descartes, *Rules for the Direction of the Mind*. Translated by E. S. Haldane and G. R. T. Ross, Encyclopaedia Britannica (Chicago, 1971).

12. René Descartes, *Discourse on Method*.

13. René Descartes, *Rules for the Direction of the Mind*.

14. *Ibid.*

15. Jacques Maritain, *The Dream of Descartes* (London, 1946).

16. *Ibid.*

17. Descartes, *Rules for the Direction of the Mind*.

18. Etienne Gilson, *The Unity of Philosophical Experience* (London, 1938).

19. *Ibid.*

20. Blaise Pascal, *Pensées*, Section II, No. 169.

21. Quoted in *Great Books of the Western World: The Great Ideas*, Vol. 1, Chapter 33 (Chicago, 1953).

22. *Summa contra Gentiles*, Vol. 1 (London, 1924–1928).

23. *Summa contra Gentiles*, Vol. 3.

Chapter 2. Levels of Being

1. Arthur O. Lovejoy, *The Great Chain of Being* (New York, 1960).

2. Catherine Roberts, *The Scientific Conscience* (Fontwell, Sussex, 1974).

Chapter 3. Progressions

1. Romans 7:14ff. Phillips translation.

2. Vol. 5, Book Four, Chapter XI.

3. Maurice Nicoll, *Living Time*, Chapter 1 (London, 1952).

4. *Ibid.*

5. By Gurdjieff to his pupils.

6. Luke 12:6.

7. Revelation 10:5–6.

Chapter 4. *"Adaequatio"*: I

1. G. N. M. Tyrrell, *Grades of Significance* (London, 1930).

2. R. L. Gregory, *Eye and Brain—The Psychology of Seeing* (London, 1966).

3. Tyrrell, *op. cit.*

4. Matthew 13:13.

5. Matthew 13:15; Acts 28:27. (Emphasis added.)

6. Revelation 3:15–16.

7. Etienne Gilson, *The Christian Philosophy of Saint Augustine* (London, 1961).

8. *Ibid.*

9. Jalal al-Din Rumi, *Mathnawi,* Vol. 4 (Gibb Memorial Series) (London, 1926–1934).

10. John Smith the Platonist, *Select Discourses* (London, 1821).

11. Richard of Saint-Victor, *Selected Writings on Contemplation* (London, 1957).

12. *Suttanipata,* IV, ix, 3.

13. *Majjhima Nikaya,* LXX.

14. John 8:32.

15. Nicoll, *op. cit.,* Chapter X (London, 1952).

Chapter 5. *"Adaequatio"*: II

1. Sir Arthur Eddington, *The Philosophy of Physical Science* (London, 1939).

2. René Descartes, Preface to the French translation of *Principia Philosophiae,* Part II.

3. *Ibid.*

4. Gilson, *The Christian Philosophy of Saint Augustine, op. cit.*

5. Matthew 19:26.

6. Gilson, *The Unity of Philosophical Experience, op. cit.*

7. *Ibid.*

8. Abraham Maslow, *The Psychology of Science,* Chapter 4 (New York, 1966). (Emphasis added.)

9. William James, *The Will to Believe* (London, 1899).

Chapter 6. The Four Fields of Knowledge: 1

1. All quotes are from Whitall N. Perry, *A Treasury of Traditional Wisdom* (London, 1971).

2. P. D. Ouspensky, *The Psychology of Man's Possible Evolution*, First Lecture (London, 1951).

3. Joseph Campbell, *The Hero with a Thousand Faces*, Prologue (New York, 1949).

4. Ernest Wood, *Yoga*, Chapter 4 (London, 1959).

5. Ouspensky, *op. cit.*

6. Nyanaponika Thera, *The Heart of Buddhist Meditation, a Handbook of Mental Training Based on the Buddha's Way of Mindfulness*, Introduction (London, 1962).

7. *Ibid.*

8. *The Instruction to Bahiya*, quoted by Nyanaponika Thera, *op. cit.*

9. *The Cloud of Unknowing*, a new translation by Clifton Wolters (London, 1961).

10. *Majjhima Nikaya*, CXL. Cf. also J. Evola, *The Doctrine of Awakening*, Chapter IV (London, 1951).

11. *The Cloud of Unknowing, op. cit.*

12. Luke 18:1. (Emphasis added.)

13. In *A Treasury of Russian Spirituality*, compiled and edited by G. P. Fedotov (London, 1952).

14. See *Writings from the Philokalia on Prayer of the Heart* (London, 1951) and *Early Fathers from the Philokalia* (London, 1954)

15. Hieromonk Kallistos (Timothy Ware) in his introduction to *The Art of Prayer* (see next note).

16. *The Art of Prayer: An Orthodox Anthology*, compiled by Igumen Chariton of Valamo, Chapter III, iii (London, 1966).

17. Wilder Penfield, *The Mystery of the Mind* (Princeton, 1975).

18. *Ibid.*

19. "On the Prayer of Jesus," from the *Ascetic Essays of Bishop Ignatius Brianchaninov* (London, 1952). The quotations are from the Introduction by Alexander d'Agapeyeff.

20. W. T. Stace, *Mysticism and Philosophy* (London, 1961).

21. *Ibid.*

22. *Ibid.* (Emphasis added.)

23. II Corinthians 4:18.

Chapter 7. The Four Fields of Knowledge: 2

1. Romans 8:22.

2. J. G. Bennett, *The Crisis in Human Affairs*, Chapter 6 (London, 1948).

3. William James, *The Principles of Psychology*, Chapter 25 (Chicago, 1952).

4. The Venerable Mahasi Sayadaw, *The Progress of Insight Through the Stages of Purification*, Chapter IV, 4 (Kandy, Ceylon, 1965).

5. St. John of the Cross, *Ascent of Mount Carmel*, Book II, Chapter 11. In *The Complete Works of Saint John of the Cross*, translated and edited by E. Allison Peers (London, 1935). (Emphasis added.)

6. Wood, *op. cit.*

7. W. Y. Evans-Wentz, *Tibetan Yoga and Secret Doctrines*, General Introduction, section XI (London, 1935).

8. *Ibid.*

9. Ernest Wood, *Practical Yoga, Ancient and Modern*, Chapter 4 (London, 1951).

10. Lorber's writings are published, in German only, by Lorber-Verlag, Bietigheim, Württemberg, West Germany.

11. See books on Edgar Cayce by Hugh Lynn Cayce and Edgar Evans Cayce (his sons), Thomas Sugrue, M. E. Penny Baker, Elsie Sechrist, W. and G. McGarey, Mary Ellen Carter, Doris Agee, Noel Langley, Harmon Hartzell Bro. (New York).

12. Doris Agee, *Edgar Cayce on ESP*, Chapter Two (New York, 1969). (Emphasis added.)

Chapter 8. The Four Fields of Knowledge: 3

1. Maurice Nicoll, *Psychological Commentaries on the Teaching of G. I. Gurdjieff and P. D. Ouspensky*, Vol. 1, p. 266 (London, 1952–1956).

2. *Ibid.*, Vol. 4, p. 1599.

3. *Ibid.*, Vol. 1, p. 267.

4. *Ibid.*, Vol. 1, p. 259.

Chapter 9. The Four Fields of Knowledge: 4

1. Arthur Livingston in his Editor's Note to *The Mind and Society* (see next note).

2. Vilfredo Pareto, *The Mind and Society,* paragraphs 69,2; 99/100; 109/110 (London, 1935).

3. *Ibid.* (Emphasis added.)

4. *Ibid.*

5. F. S. C. Northrop, *The Logic of the Sciences and the Humanities,* Chapter 8 (New York, 1959). (Emphasis added.)

6. René Guénon, *Symbolism of the Cross,* Chapter IV (London, 1958).

7. "Nature, Philosophy of," in *The New Encyclopaedia Britannica* (1975), Vol. 12, p. 873.

8. *Ibid.*

9. Karl Stern, *The Flight from Woman,* Chapter 5 (New York, 1965).

10. René Descartes, *Discourse on Method,* Part VI.

11. Julian Huxley, *Evolution: The Modern Synthesis,* Chapter 2, section 7 (London, 1942).

12. "Evolution," in *The New Encyclopaedia Britannica,* Vol. 7, pp. 23 and 17.

13. *Ibid.*

14. *Ibid.*

15. *Ibid.*

16. Stern, *op. cit.,* Chapter 12.

17. The *Times* (London) reports on January 24, 1977: "Mr. John Watson, head of the religious education department of Rickmansworth School, who was dismissed for teaching the literal 'Genesis' view of creation instead of the evolutionary view favoured in the agreed syllabus, intends . . . [to take legal action]. Mr. Watson . . . was a missionary in India for 16 years and is the author of two books that put forward the Genesis theory of creation."

A "Monkey Trial in reverse," to show that all *Faiths* tend to be intolerant!

18. *Op. cit.*

19. Douglas Dewar, *The Transformist Illusion* (Dehoff Publications, Murfreesboro, Tennessee, 1957).

20. Martin Lings in *Studies in Comparative Religion*, Vol. 4, No. 1, 1970, p. 59; published quarterly by Tomorrow Publications Ltd., Bedfont, Middlesex, England.

21. Harold Saxton Burr, *Blueprint for Immortality: The Electric Patterns of Life* (London, 1972).

22. *Ibid.* (Emphasis added.)

23. *Ibid.*

Chapter 10. Two Types of Problems

1. See René Guénon, *The Reign of Quantity and the Signs of the Times,* translated by Lord Northbourne (London, 1953).

2. Cf. Paul Roubiczek, *Existentialism . . . For and Against* (Cambridge, 1964).

3. Thomas Aquinas, Commentary on the Gospel of Matthew 5:2.

4. Ananda K. Coomaraswamy, *Christian and Oriental Philosophy of Art,* Chapter One, "Why Exhibit Works of Art?" (New York, 1956). (Emphasis added.)

5. Quoted in Dorothy L. Sayers, *Further Papers on Dante,* p. 54 (London, 1957).

6. Dante, *The Divine Comedy,* translated by Charles Eliot Norton, Great Books of the Western World, Encyclopaedia Britannica (Chicago, 1952).

7. P. B. Medawar, *The Art of the Soluble,* Introduction (London, 1967).

Epilogue

1. Dorothy L. Sayers, *Introductory Papers on Dante,* p. 114 (London, 1954).

About the author

About the book

Insights,
Interviews
& More . . .

Read on

Meet E. F. Schumacher

Dr. E. F. Schumacher (1911–1977) was born in Bonn, Germany. A Rhodes scholar in economics at the University of Oxford, Schumacher moved to the United States in 1933 to teach at Columbia University. Unsatisfied teaching theoretical economics without practical applications, Schumacher returned to Germany, working as a businessman, farmer, and journalist. But as Hitler's Third Reich increasingly transformed Germany, Schumacher found life there untenable and made a final move back to England in 1937. There, he exchanged ideas with, and helped influence, John Maynard Keynes. During Schumacher's career in England, he held various important posts, including chief economist for Britain's National Coal Board, president of the Soil Association, and founder of Practical Action, an international nongovernmental organization that fights poverty in the developing world.

Schumacher remains best known for the two books he published during his lifetime: *Small Is Beautiful: Economics as if People Mattered* and *A Guide for the Perplexed*. In them, he lays out the critique of western economics that defined his philosophy and life, explaining why a modern economy focused on growth at all costs has had disastrous consequences, and offering a sustainable alternative. Schumacher's philosophy has been enormously influential for a generation of economists, environmentalists, and activists. The Schumacher Center for a New Economics, based in Massachusetts, continues his work in the twenty-first century. ∾

Buddhist Economics
An Excerpt from
Small Is Beautiful

In this passage, excerpted from Small Is Beautiful, *E. F. Schumacher explores the modern economy through a Buddhist lens.*

"RIGHT LIVELIHOOD" is one of the requirements of the Buddha's Noble Eightfold Path. It is clear, therefore, that there must be such a thing as Buddhist economics.

Buddhist countries have often stated that they wish to remain faithful to their heritage. So Burma: "The New Burma sees no conflict between religious values and economic progress. Spiritual health and material well-being are not enemies: they are natural allies." Or: "We can blend successfully the religious and spiritual values of our heritage with the benefits of modern technology." Or: "We Burmans have a sacred duty to conform both our dreams and our acts to our faith. This we shall ever do."

All the same, such countries invariably assume that they can model their economic development plans in accordance with modern economics, and they call upon modern economists from so-called advanced countries to advise them, to formulate the policies to be pursued, and to construct the grand design for development, the Five-Year Plan or whatever it may be called. No one seems to think that a Buddhist way of life would call for Buddhist economics, just as the modern materialist way of life has brought forth modern economics.

Economists themselves, like most specialists, normally suffer from a kind of metaphysical blindness, assuming that theirs is a science of absolute and invariable truths, without any presuppositions. Some go as far as to claim that economic laws are as free from "metaphysics" or "values" as the law of gravitation. We need not, however, get involved in arguments of methodology. Instead, let us take some fundamentals and see what they look like when viewed by a modern economist and a Buddhist economist.

There is universal agreement that a fundamental source of wealth is human labour. Now, the modern economist has been brought up to consider "labour" or work as little more than a necessary evil. From the point of view of the employer, it is in any case simply an item of cost, to be reduced to a minimum if it cannot be eliminated altogether, say, by automation. From the point of view of the workman, it is a "disutility"; to work is to make a sacrifice of one's leisure and comfort, and wages are a kind of compensation for the sacrifice. Hence the ideal from the point of view of the employer is to have output without employees, and the ideal from the point of view of the employee is to have income without employment.

The consequences of these attitudes both in theory and in practice are, of course, extremely far-reaching. If the ideal with regard to work is to get rid ▶

Buddhist Economics *(continued)*

of it, every method that "reduces the work load" is a good thing. The most potent method, short of automation, is the so-called "division of labour" and the classical example is the pin factory eulogised in Adam Smith's *Wealth of Nations*. Here it is not a matter of ordinary specialisation, which mankind has practised from time immemorial, but of dividing up every complete process of production into minute parts, so that the final product can be produced at great speed without anyone having had to contribute more than a totally insignificant and, in most cases, unskilled movement of his limbs.

The Buddhist point of view takes the function of work to be at least threefold: to give a man a chance to utilise and develop his faculties; to enable him to overcome his ego-centredness by joining with other people in a common task; and to bring forth the goods and services needed for a becoming existence. Again, the consequences that flow from this view are endless. To organise work in such a manner that it becomes meaningless, boring, stultifying, or nerve-racking for the worker would be little short of criminal; it would indicate a greater concern with goods than with people, an evil lack of compassion, and a soul-destroying degree of attachment to the most primitive side of this worldly existence. Equally, to strive for leisure as an alternative to work would be considered a complete misunderstanding of one of the basic truths of human existence, namely that work and leisure are complementary parts of the same living process and cannot be separated without destroying the joy of work and the bliss of leisure.

From the Buddhist point of view, there are therefore two types of mechanisation which must be clearly distinguished: one that enhances a man's skill and power and one that turns the work of man over to a mechanical slave, leaving man in a position of having to serve the slave. How to tell the one from the other? "The craftsman himself," says Ananda Coomaraswamy, a man equally competent to talk about the modern West as the ancient East, "can always, if allowed to, draw the delicate distinction between the machine and the tool. The carpet loom is a tool, a contrivance for holding warp threads at a stretch for the pile to be woven round them by the craftsmen's fingers; but the power loom is a machine, and its significance as a destroyer of culture lies in the fact that it does the essentially human part of the work." It is clear, therefore, that Buddhist economics must be very different from the economics of modern materialism, since the Buddhist sees the essence of civilisation not in a multiplication of wants but in the purification of human character. Character, at the same time, is formed primarily by a man's work. And work, properly conducted in conditions of human dignity and freedom, blesses those who do it and equally their ▶

products. The Indian philosopher and economist J. C. Kumarappa sums the matter up as follows:

> If the nature of the work is properly appreciated and applied, it will stand in the same relation to the higher faculties as food is to the physical body. It nourishes and enlivens the higher man and urges him to produce the best he is capable of. It directs his free will along the proper course and disciplines the animal in him into progressive channels. It furnishes an excellent background for man to display his scale of values and develop his personality.

If a man has no chance of obtaining work he is in a desperate position, not simply because he lacks an income but because he lacks this nourishing and enlivening factor of disciplined work which nothing can replace. A modern economist may engage in highly sophisticated calculations on whether full employment "pays" or whether it might be more "economic" to run an economy at less than full employment so as to ensure a greater mobility of labour, a better stability of wages, and so forth. His fundamental criterion of success is simply the total quantity of goods produced during a given period of time. "If the marginal urgency of goods is low," says Professor Galbraith in *The Affluent Society*, "then so is the urgency of employing the last man or the last million men in the labour force."

And again: "If . . . we can afford some unemployment in the interest of stability—a proposition, incidentally, of impeccably conservative antecedents— then we can afford to give those who are unemployed the goods that enable them to sustain their accustomed standard of living."

From a Buddhist point of view, this is standing the truth on its head by considering goods as more important than people and consumption as more important than creative activity. It means shifting the emphasis from the worker to the product of work, that is, from the human to the subhuman, a surrender to the forces of evil. The very start of Buddhist economic planning would be a planning for full employment, and the primary purpose of this would in fact be employment for everyone who needs an "outside" job: it would not be the maximisation of employment nor the maximisation of production. Women, on the whole, do not need an "outside" job, and the large-scale employment of women in offices or factories would be considered a sign of serious economic failure. In particular, to let mothers of young children work in factories while the children run wild would be as uneconomic in the eyes of a Buddhist economist as the employment of a skilled worker as a soldier in the eyes of a modern economist.

While the materialist is mainly ▶

interested in goods, the Buddhist is mainly interested in liberation. But Buddhism is "The Middle Way" and therefore in no way antagonistic to physical well-being. It is not wealth that stands in the way of liberation but the attachment to wealth; not the enjoyment of pleasurable things but the craving for them. The keynote of Buddhist economics, therefore, is simplicity and nonviolence. From an economist's point of view, the marvel of the Buddhist way of life is the utter rationality of its pattern—amazingly small means leading to extraordinarily satisfactory results.

For the modern economist this is very difficult to understand. He is used to measuring the "standard of living" by the amount of annual consumption, assuming all the time that a man who consumes more is "better off" than a man who consumes less. A Buddhist economist would consider this approach excessively irrational: since consumption is merely a means to human well-being, the aim should be to obtain the maximum of well-being with the minimum of consumption. Thus, if the purpose of clothing is a certain amount of temperature comfort and an attractive appearance, the task is to attain this purpose with the smallest possible effort, that is, with the smallest annual destruction of cloth and with the help of designs that involve the smallest possible input of toil. The less toil there is, the more time and strength is left for artistic creativity. It would be highly uneconomic, for instance, to go in for complicated tailoring like the modern West, when a much more beautiful effect can be achieved by the skilful draping of uncut material. It would be the height of folly to make material so that it should wear out quickly and the height of barbarity to make anything ugly, shabby or mean. What has just been said about clothing applies equally to all other human requirements. The ownership and the consumption of goods is a means to an end, and Buddhist economics is the systematic study of how to attain given ends with the minimum means.

Modern economics, on the other hand, considers consumption to be the sole end and purpose of all economic activity, taking the factors of production—land, labour, and capital— as the means. The former, in short, tries to maximise human satisfactions by the optimal pattern of consumption, while the latter tries to maximise consumption by the optimal pattern of productive effort. It is easy to see that the effort needed to sustain a way of life which seeks to attain the optimal pattern of consumption is likely to be much smaller than the effort needed to sustain a drive for maximum consumption. We need not be surprised, therefore, that the pressure and strain of living is very much less in, say, Burma than it is in the United States, in spite of the fact that the amount of labour-saving machinery used in the former country ▶

is only a minute fraction of the amount used in the latter.

Simplicity and nonviolence are obviously closely related. The optimal pattern of consumption, producing a high degree of human satisfaction by means of a relatively low rate of consumption, allows people to live without great pressure and strain and to fulfil the primary injunction of Buddhist teaching: "Cease to do evil; try to do good." As physical resources are everywhere limited, people satisfying their needs by means of a modest use of resources are obviously less likely to be at each other's throats than people depending upon a high rate of use. Equally, people who live in highly self-sufficient local communities are less likely to get involved in large-scale violence than people whose existence depends on worldwide systems of trade.

From the point of view of Buddhist economics, therefore, production from local resources for local needs is the most rational way of economic life, while dependence on imports from afar and the consequent need to produce for export to unknown and distant peoples is highly uneconomic and justifiable only in exceptional cases and on a small scale. Just as the modern economist would admit that a high rate of consumption of transport services between a man's home and his place of work signifies a misfortune and not a high standard of life, so the Buddhist economist would hold that to satisfy human wants from faraway sources rather than from sources nearby signifies failure rather than success. The former tends to take statistics showing an increase in the number of tons/miles per head of the population carried by a country's transport system as proof of economic progress, while to the latter—the Buddhist economist—the same statistics would indicate a highly undesirable deterioration in the *pattern* of consumption.

Another striking difference between modern economics and Buddhist economics arises over the use of natural resources. Bertrand de Jouvenel, the eminent French political philosopher, has characterised "Western man" in words which may be taken as a fair description of the modern economist:

He tends to count nothing as an expenditure, other than human effort; he does not seem to mind how much mineral matter he wastes and, far worse, how much living matter he destroys. He does not seem to realise at all that human life is a dependent part of an ecosystem of many different forms of life. As the world is ruled from towns where men are cut off from any form of life other than human, the feeling of belonging to an ecosystem is not revived. This results in a harsh and improvident treatment of things upon which ▶

we ultimately depend, such as water and trees.

The teaching of the Buddha, on the other hand, enjoins a reverent and nonviolent attitude not only to all sentient beings but also, with great emphasis, to trees. Every follower of the Buddha ought to plant a tree every few years and look after it until it is safely established, and the Buddhist economist can demonstrate without difficulty that the universal observation of this rule would result in a high rate of genuine economic development independent of any foreign aid. Much of the economic decay of southeast Asia (as of many other parts of the world) is undoubtedly due to a heedless and shameful neglect of trees.

Modern economics does not distinguish between renewable and nonrenewable materials, as its very method is to equalise and quantify everything by means of a money price. Thus, taking various alternative fuels, like coal, oil, wood, or waterpower: the only difference between them recognised by modern economics is relative cost per equivalent unit. The cheapest is automatically the one to be preferred, as to do otherwise would be irrational and "uneconomic." From a Buddhist point of view, of course, this will not do; the essential difference between nonrenewable fuels like coal and oil on the one hand and renewable fuels like wood and waterpower on the other cannot be simply overlooked.

Nonrenewable goods must be used only if they are indispensable, and then only with the greatest care and the most meticulous concern for conservation. To use them heedlessly or extravagantly is an act of violence, and while complete nonviolence may not be attainable on this earth, there is nonetheless an ineluctable duty on man to aim at the ideal of nonviolence in all he does.

Just as a modern European economist would not consider it a great economic achievement if all European art treasures were sold to America at attractive prices, so the Buddhist economist would insist that a population basing its economic life on nonrenewable fuels is living parasitically, on capital instead of income. Such a way of life could have no permanence and could therefore be justified only as a purely temporary expedient. As the world's resources of nonrenewable fuels—coal, oil and natural gas—are exceedingly unevenly distributed over the globe and undoubtedly limited in quantity, it is clear that their exploitation at an ever-increasing rate is an act of violence against nature which must almost inevitably lead to violence between men.

This fact alone might give food for thought even to those people in Buddhist countries who care nothing for the religious and spiritual values of their heritage and ardently desire to embrace the materialism of modern economics at the fastest possible speed. Before they ▶

dismiss Buddhist economics as nothing better than a nostalgic dream, they might wish to consider whether the path of economic development outlined by modern economics is likely to lead them to places where they really want to be. Towards the end of his courageous book *The Challenge of Man's Future*, Professor Harrison Brown of the California Institute of Technology gives the following appraisal:

> Thus we see that, just as industrial society is fundamentally unstable and subject to reversion to agrarian existence, so within it the conditions which offer individual freedom are unstable in their ability to avoid the conditions which impose rigid organisation and totalitarian control. Indeed, when we examine all of the foreseeable difficulties which threaten the survival of industrial civilisation, it is difficult to see how the achievement of stability and the maintenance of individual liberty can be made compatible.

> Even if this were dismissed as a long-term view there is the immediate question of whether "modernisation," as currently practised without regard to religious and spiritual values is actually producing agreeable results. As far as the masses are concerned, the results appear to be disastrous—a collapse of the rural economy, a rising tide of unemployment

in town and country, and the growth of a city proletariat without nourishment for either body or soul.

It is in the light of both immediate experience and long-term prospects that the study of Buddhist economics could be recommended even to those who believe that economic growth is more important than any spiritual or religious values. For it is not a question of choosing between "modern growth" and "traditional stagnation." It is a question of finding the right path of development, the Middle Way between materialist heedlessness and traditionalist immobility, in short, of finding "Right Livelihood." ☙

Have You Read?
More by
E. F. Schumacher

SMALL IS BEAUTIFUL

Small Is Beautiful is E. F. Schumacher's classic call for the end of excessive consumption. Schumacher inspired such movements as "Buy Locally" and "Fair Trade," while voicing strong opposition to wasteful corporate behemoths. Named one of the *Times Literary Supplement*'s 100 Most Influential Books Since World War II, *Small Is Beautiful* presents eminently logical arguments for building our economies around the needs of communities, not corporations. This reissue of Schumacher's "eco-bible" (*Time*), featuring a foreword by leading environmental activist Bill McKibben, is as relevant now as it has ever been.

"A masterpiece. . . . Embracing what Schumacher stood for—above all the idea of sensible scale—is the task for our time. *Small Is Beautiful* could not be more relevant."
> —Bill McKibben, from the foreword

"Nothing less than a full-scale assault on conventional economic wisdom."
> —*Newsweek*

Discover great authors, exclusive offers, and more at hc.com.

"*Small Is Beautiful* changed the way many people think about bigness and its human costs." —*New York Times*